A PRACTICAL GUIDE FOR
BUSINESS SPEAKING

A Practical Guide for Business Speaking

EDWARD P. BAILEY, JR.

New York   Oxford
OXFORD UNIVERSITY PRESS
1992

Oxford University Press

Oxford   New York   Toronto
Delhi   Bombay   Calcutta   Madras   Karachi
Kuala Lumpur   Singapore   Hong Kong   Tokyo
Nairobi   Dar es Salaam   Cape Town
Melbourne   Auckland

and associated companies in
Berlin   Ibadan

Published by Oxford University Press, Inc.
200 Madison Avenue, New York, NY 10016

Oxford is a registered trademark of Oxford University Press

Library of Congress Cataloging-in-Publication Data
Bailey, Edward P.
A practical guide for business speaking
Edward P. Bailey, Jr.
p.   cm.
Includes index
ISBN 0-19-507361-4
1. Business presentations I. Title.
HF5718.22.B35 1992
658.4'52—dc20   92-4898

2 4 6 8 9 7 5 3 1

Printed in the United States of America
on acid-free paper

For my wife, Janet,
and parents, Edward and Marilou

Acknowledgments

Two of my colleagues have greatly influenced this book—Dr. Tom Murawski and Dr. Terry Bangs:

- Tom helped me get started as a speaker. He stands out for his captivating speaking style—focused, creative, energetic. Those who have seen him speak remember him always.

- Terry's also a masterful speaker. I remember watching him speak to 800 people in an auditorium, leaning back against a chair, arms folded, simply talking as though only two or three others were in the room. His natural manner held the audience spellbound.

My sincere thanks to both of them for their friendship and the model they set for me.

Other people have been very helpful:

- Janet Hiller (my wife and co-consultant) who read and commented on everything *and* completely typeset the entire book. She is terrific.

- Brooke Bailey (my brother and co-consultant) who gave generously of his time and energy. He's an excellent speaker, too!

- Dr. Bob Sigethy, Professors Joan Feeney and Maribeth Wyvill, and Sister P. J. Cahill—my colleagues at Marymount University.

- Don Insko, an expert in business presentation graphics, for his comments and important contributions to the graphical elements of the book.

- Phil Powell—longstanding friend, co-author on three books, and "world-class" editor.

- Dr. Jim Gaston—friend and colleague for nearly twenty years.

- Dr. Fred Kiley, Dr. Greg Foster, and Ms. Judy Clark—my friends at the National Defense University.

- Marilou and Edward Bailey, my parents; and Jeannette and Laura Bailey, my daughters. They provided helpful comments and inspiration.

My thanks to speakers whose presentations helped generate ideas for this book: Dr. Robert Anthony, Michael Gallagher, Brenda Jones, Michael Kopito, Sheila Marion, Doyle Mitchell, Guy Sahatjian, Deborah Tompkins-Lipscomb, and hundreds of others throughout the years.

And also my thanks to my other reviewers for their helpful insight: Max Boot, Bob Brofft, and Jim Casimir.

My acknowledgment to the Coca-Cola Company for granting me permission to use its trademark.

Finally, my appreciation to my editors at Oxford University Press: Liz Maguire, Ellen Fuchs, and Susan Chang. I am indeed fortunate to have worked with Oxford and with them.

Arlington, Va.   E. P. B.
March 1992

Contents

APPENDIXES

DESIGNING YOUR
PRESENTATION

CHAPTER 1

Overview

Have you ever sat through a presentation wondering, "What is this thing about? And when is it *ever* going to end?"

We've all been there, haven't we? This book may not help you as a listener, but it will help you as a speaker—so you aren't the one causing the mental grumbling in your audience.

You don't need to be a "born speaker" to speak well. Very few people are. Many people who seem to be born speakers have actually gotten there through practice, more practice, and lots of experience.

But that practice isn't haphazard.

This book suggests an approach for preparing your presentations, an approach I almost always use. And this approach is practical, emphasizing *how* rather than *why*:

- how to organize your presentation
- how to remember what you plan to say
- how to design visual aids
- how to rehearse
- how to prepare the room for your presentation
- how to handle questions and answers

And more.

What is a "business" presentation?

There is no standard business presentation. Perhaps the speaker is standing in front of a small group of people talking about work. Or selling them something. Or urging them to make a change.

Sometimes a business presentation is aimed at co-workers. Or bosses. Or subordinates. Or clients.

Sometimes it takes place in someone's office. Or in a small conference room. Or in a hotel auditorium with hundreds of people in the audience—the speaker hooked up to a microphone, visual aids staff at the ready, lectern furnished with a pitcher of cold water (and the speaker's hands even colder than that).

This book will help with the entire range of business presentations—because the fundamentals are the same.

What is the book's main message?

This is the main message: most presentations succeed or fail long before the speaker stands in front of the audience. Most presentations succeed or fail in the design stage—because a good presentation is *designed*.

By "designed" I don't mean something the graphics department does. I mean something the speaker does: the thoughtful, meticulous, purposeful preparation that helps the speaker communicate. Just as important, that preparation makes the speech far, far easier to give.

A good design, in other words, is the best way to take the pressure off. Many inexperienced speakers design a presentation that's almost bound to fail. This book will help you design one that's almost bound to succeed. Well . . . *likely* to succeed.

What, specifically, does this book cover?

This book sets forth a process for designing and giving your presentation:

- *Organizing your presentation.* You'll want to begin by roughing out the general structure of your presentation—making sure it has an absolutely clear organization. If not, listeners will probably get lost, and listeners who get lost rarely find their way back. The good news is there's an organization—a pattern—that works for most business presentations.

- *Using examples.* Many presentations depend on the audience's understanding a new term—things like radiometry, type amendment, live-fire vulnerability testing, upselling, passive activity, and so on. How can you be sure you communicate your key terms? The answer: examples. In fact, a well-placed example can be the difference in whether your audience even knows what you're talking about.

- *Remembering what you plan to say.* This is where most presentations get derailed. Once you've decided what to say, you need to remember it and there are various ways to do that. Memorizing is the worst way. Well-designed visual aids are often the best.

- *Choosing visual aids.* Most business presentations rightly depend on visual aids. But when should you use an overhead projector? A flip chart? A computer presentation? Thirty-five-millimeter slides? You need to know the advantages and disadvantages of each.

- *Designing visual aids.* What should you keep in mind as you design your visual aids? How much material you put on them is extremely important. Put too little on, and you and your audience may get lost. Put too much on, and you end up simply reading aloud. Put the words in the wrong place, and people in the back of the room may not be able to see them.

- *Using audience participation and humor.* Most good presentations have energy flowing through them. There are ways to involve your audience even in formal speaking situations. There are also pitfalls, especially with humor, that you must be aware of.

- *Rehearsing.* Once you have the content set for your presentation, you need to rehearse. Rehearsing just means standing and saying the words to an empty room, right? To some extent, but there are other things you can do to make your rehearsing more efficient. Good speakers often spend a lot of time rehearsing.

- *Setting up the room.* If the room is too hot or people can't see your visual aids, you're not going to communicate well. Sometimes you don't have much control of the room, but most speakers have more control than they suspect. I'll cover what to look for in each room and how to keep problems from happening. There's a real art to this.

- *Using effective techniques of delivery.* Even with a perfect design, a presentation will fail if the speaker qualifies for *The Guinness Book of Records* for nervous pacing and "uh's" per second. There are important do's and don't's you should know.

- *Presenting visual aids.* The best visual aids in the world are worthless unless you know how to present them to your audience: where to stand, how to use a pointer, when to read your visual aids aloud, and when not to.

- *Handling questions and answers.* What happens when your formal presentation is over? Are you out of the woods yet? Most of the time you're not: there's a question and answer session. You can prepare for this, too. It can even be the most powerful part of your presentation.

- *Helping others speak better.* I'll finish with some advice on helping your co-workers be better speakers.

This book has three appendixes: one is a checklist for speakers; another is a checklist for setting up the room; and the last shows a good way to organize your presentation.

Does all this sound like a lot of work? Well, "effortless" presentations are usually the product of a lot of effort. But the more you understand about presentations, the more comfortable you'll be in preparing them and—more important—in giving them.

That is, the more you feel in control, the more confident you'll be when the spotlight goes on . . . and you're in it.

So let's get started!

CHAPTER 2

Organizing Your Presentation

When you're reading a book and get lost, you can flip back and start over. But when you're listening to a speech and get lost, well . . . there's not much you can do except sit there and consult your watch, the window, and the interior of your mind.

So when I prepare a presentation, I try to make its organization absolutely clear. I think of a presentation as a trip, and I don't want my audience to get lost. I want them to know, at the beginning, where we're going and how we'll get there. And when the trip is underway, I want them to know exactly where we are—what town we've just passed and what's coming up.

So here's what I suggest:

- Announce your topic, define any terms in it, and state your bottom line.
- Then outline for your audience what the major parts of your presentation will be—that's called your *blueprint.*
- And plan strong, obvious transitions throughout.

That said, there is no one way—no one organizational pattern—to communicate all ideas to all people. But I find this approach works most of the time. I use it when I first start to prepare. It serves me well as a starting point, and it often becomes the final organization I use.

Let's look at these suggestions more closely.

Announce your topic, define terms, and state your bottom line

My first suggestion is to let members of your audience know the destination for their "trip." To do that, you'll need to announce your topic, define unfamiliar terms in it, and state your bottom line up front.

Announce your topic

Announcing your topic is simple: just say what you're going to talk about. You don't have to make the topic the first words out of your mouth. You might want to say a few polite sentences first—what linguists call polite noise: "Thank you, Jeannette, for. . . ." And a little humor up front lets groups settle in and become willing to listen. But don't wait long.

To announce your topic, simply say something like:

- My purpose is to tell you how our company is doing financially.
- I'm here to talk about the promotion system for our junior staff.
- This afternoon I'm going suggest a way to improve the quality of our proposals.

Those are so straightforward, aren't they?—clean, simple, efficient starts. The audience should begin to relax, just knowing that you actually have a clear purpose and are able to articulate it.

Define unfamiliar terms in your topic

Sometimes the topic isn't so straightforward. There may be a term in it, a necessary one, that's unfamiliar to your audience.

Here are some examples:

- Today, I'll explain the value of *type amendment.*
- I'm going to demonstrate a new *model of unit strength and cohesion.*
- I'll suggest that *fractals* can describe the behavior of two companies competing with each other.

If the topic itself has unfamiliar words in it, your audience needs to understand them immediately. Yet you've probably seen speakers go into the advantages of something like "type amendment" before defining it.

So right after you announce your topic, define any unfamiliar words. For example, you could explain that *type amendment* means "changing the shapes of letters in the alphabet ('type-faces') to make them visually interesting or memorable."

But that still doesn't really communicate, does it?

What the audience needs now is a quick example: "Let me give you an example. In the words 'Coca-Cola,' each capital *C* has a unique shape—extending into a long, curvy line. That's one example of type amendment. The shape of the letter *C* is changed to make it visually interesting—and memorable."

Even better, use a visual aid. The next page shows you the words "Coca Cola."

To define your term, then, normally give a short plain English definition and follow with a quick example. The little time this takes is absolutely crucial for your audience.

Type Amendment

Trade-mark ®

Tip

Define only your key term in your introduction—only what your audience must know right away. Save other definitions for just when your audience needs them.

State your bottom line

Now for what audiences most want to hear up front: your bottom line. Rudolf Flesch, a pioneer in plain English writing, called that "spilling the beans." Spilling the beans is even more crucial for speaking than it is for writing. In other words, make your bottom line your top line by telling your audience, right up front, your conclusion, recommendation, or request.

For example:

- Overall, you'll be pleased with our company's financial picture. Our profits are up 8% and our predictions are all good. Now let's look at the details. [a conclusion]

- The promotion system for our junior staff isn't working. I'll explain the problems and suggest we set up a promotion board to make all selections. [a recommendation]

- Our division needs a color printer to help us prepare more professional proposals. [a request]

Some speakers don't like this approach, hating to give away the bottom line at the beginning. They feel audiences are more receptive who first hear *all* the facts and *all* the logical arguments and then hear the bottom line at the end.

However, I don't think their audiences would agree. From my experience, audiences that have to wait get impatient. Worse, they often get confused: all those facts and arguments simply don't mean much without the essential context of the bottom line. Imagine yourself as part of the audience: when would you want the speaker to tell you the bottom line?

So get to the point—right up front. What works best for you as a listener also works best for your audience.

Tip

When you state your bottom line up front, be sure you don't give only your topic. Actually state, briefly, what you conclude, recommend, or request.

Sometimes a presentation doesn't have a bottom line. In that case, be sure your audience isn't expecting one. That is, tell them your purpose: you're simply providing information and not asking for any action. Here's an example of a poor topic statement that may or may not have a bottom line later on:

- The topic today is our benefits program.

Is the speaker asking for more money for benefits? Or just providing information? It's impossible to tell.

This topic statement makes it clearer there's no delayed bottom line:

- This morning I'll update you on our costs last month for our benefits program.

Now the decision maker—and the rest of the audience—clearly knows the presentation's purpose. They aren't waiting for you to spring a money request on them.

Use a blueprint

My second suggestion is to let your audience know what subtopics are coming. In other words, let them know the various "places" you'll pass through on your "trip." The way to do that is with a blueprint. A blueprint is simply a list of the sections of your presentation.

Suppose you're a headhunter—someone who makes money by finding experienced people to move from one company to another. You're explaining to your new employees just what headhunters do. For a blueprint, you might say this: "I'll explain our three main tasks: finding job openings, finding employees, and matching employees with those openings."

This technique—telling what you're about to cover—helps the audience stay on the same track you're on. Even though it's simple, it can go wrong. Here are some suggestions.

Give a blueprint for the body only

Don't confuse the outline of your presentation with a blueprint. A blueprint covers only the points in the *body* of the presentation. All too often, I've seen speakers use something ineffective like this as the visual aid for their blueprint:

```
Outline
───────────────────────────────────────

              I. introduction
              II. finding job openings
              III. finding prospective employees
              IV. matching emloyees and openings
              V. conclusion
```

When the speaker turns to this visual aid, he finds himself saying something like this, "First, I'll give you an introduction. . . ." But he's already giving the introduction! Then he'll continue, "And then I'll cover how to find job openings, how to find employees, how to match the two, and then I'll finish with some conclusions."

To avoid these problems, the visual aid would be better like this:

```
What Headhunters Do
───────────────────────────────────────

              • find job openings

              • find prospective employees

              • match employees and openings
```

So don't think of a blueprint as an outline of your entire presentation. The blueprint outlines only the body.

Tip

Be sure the order of the items in your blueprint is the same order as the sections of your presentation. And make sure your blueprint comes JUST before you start into your first section. Delays between a blueprint and the body of a presentation usually confuse the audience.

Avoid jargon in your blueprint

Jargon is the special vocabulary of a group of people. For people who share the jargon, there's no problem. But for those who don't, it can end communication.

Try to avoid inappropriate jargon in your blueprint. Too often, we see blueprints that really don't communicate much. Suppose the topic is buying computers. This wouldn't be an effective blueprint for people who aren't especially familiar with computers, would it?

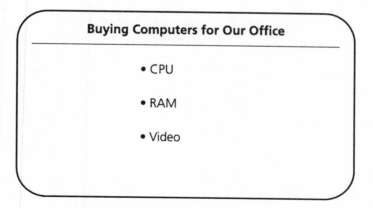

Buying Computers for Our Office

- CPU

- RAM

- Video

If people are unfamiliar with computers, the blueprint doesn't say much more than "this presentation is going to talk about three something or others."

This would be a better approach:

Buying Computers for Our Office

- How fast should they be?

- How much memory should they have?

- What type of monitor should we get?

Memory and *monitor* are familiar to most people today—certainly more familiar than *RAM* and *video*.

Don't always use a blueprint

A long or complicated blueprint can be more hindrance than help. For example, if your presentation has eight sections, a blueprint would seem awfully long. The audience could well tune you out rather than listen to all eight items. Instead, simply use an implied blueprint: "I'll cover the eight retirement plans you can choose from." You haven't named the plans, but you have told your audience the structure you plan to follow.

Use strong transitions throughout

My third suggestion is to use strong transitions and plan them as you rehearse.

We know that good organization helps. So does a good blueprint. But all may be for nothing if listeners can't figure out when you've moved from one section of your presentation to the next.

In business writing, headings signal a new topic is coming up. In speaking, you must use other techniques. These are the three techniques I suggest: announce the transition explicitly, use visual aids, and use body movement.

Announce the transition explicitly

Don't be afraid to be absolutely mechanical with your transitions. Subtlety may be nice in James Joyce's novels, but it's of little value in business presentations.

For an example, let's return to our topic on headhunters. This could be your first major transition:

- The first task for a headhunter is to find job openings to fill.

Then say what you have to say. When you're ready for the next item in your blueprint, say something like this:

- Now that we've found a job opening, we need to find someone to fill it. That brings me to our second task as headhunters.

Unmistakable. That's the "looking backward/looking forward" transition: it looks back at the section you just finished (finding jobs) and looks forward to the section you're about to start (finding employees).

This technique has a terrific advantage: it lets your audience know you're moving from one major point to another. Simply saying "next we'll look at . . ." would have been ambiguous: Were you moving from one *major* point to the next? Or were you moving from one *sub*point to the next?

With the "looking backward/looking forward" technique, your transition is clear. Your listeners know unmistakably that they've left one "town" and gone to the next.

Use visual aids to reinforce transitions

Explicitly announcing the transition is a good start, but you can reinforce your transition even more by using visual aids.

Suppose you use a visual aid near the beginning of your presentation to give your blueprint. You can use it again to move to each major point. That way, not only will your audience hear that you're making a transition—they can see it, too. You might even use highlighting on the visual aid to show the audience which point you're moving to:

What Headhunters Do

- find job openings

→ **FIND PROSPECTIVE EMPLOYEES**

- match employees and openings

Now it's easy for everyone to see what your next point is.

Use body movement to reinforce transitions

Finally, you can use your entire body to help reinforce a transition. If you're using a visual aid, move toward it and point to the next topic you're going to cover. If you're not using a visual aid, simply take a few steps to the side as you're announcing the next topic. Watch professional speakers. They often walk or make some other movement at key transition points.

To some, an obvious organization may seem terribly mechanical, but I think of it as wonderfully clear. And a clear organization doesn't squeeze the personality out of a presentation.

I've heard a fireman describe his first unsuccessful rescue—with a clear organization and tears in his listeners' eyes. I've heard a scientist describe a complex innovation for super-computers—with a clear organization and the total attention of fellow scientists and laypeople alike. And I've heard a top general in the military update his senior officers—with a clear organization and the respectful, almost affectionate, response of his people.

So don't think of a clear organization as too mechanical. Think of it as a terrific way to communicate.

Appendix C, "Model for Organizing Your Presentation," summarizes this chapter in the form of a brief model. You may want to look at it now.

CHAPTER 3

Using Examples

The last chapter gave you a good structure for a presentation—now let's talk about content. I can't help you much with your particular topic, but I can point out where the content of presentations usually goes wrong: it's too abstract. Far too often, presentations fail because the speaker uses generalizations but no examples.

Generalizations alone rarely communicate. In fact, there's some interesting research on this point. Researchers had people read abstract writing and try to figure it out. While reading, the people expressed their thoughts aloud ("Hmmm. Wonder what that means. . . ."). That way, the researchers could observe the thinking process.

Guess what happened? When the people came across an abstract idea, they tried to think of an example. In other words, they tried to make the meaning clear by turning abstract statements they didn't understand into concrete ones they did.

As speakers, we're better off providing those examples ourselves. Examples take time, but not necessarily a lot. And there's no substitute for them when it comes to communication.

Tom Murawski, a close friend and respected communication consultant, once said to me, "The two most common transitions in the language are *however* and *therefore*, but the most important is *for example.* "

Tip

As you design your presentation, check yourself using Tom's advice: how often are you saying "for example" and "for instance?" And listen to other speakers, too. You'll notice the ones who use examples and the ones who don't.

This chapter will follow Tom's advice. It's composed almost entirely of examples, all from presentations I've seen. The examples were effective for me—I still remember them!

First we'll look at quick examples and then turn to longer ones.

Quick examples

Examples don't have to take more than a few seconds. Here are several quick ones:

- Applications are simply software programs—*like word processors, spreadsheets, or even computer bridge games.*

- Our department could save money if we stopped buying individual software packages and bought network versions instead. *For example, word processing software cost us about $5,000 last year. But the network version of the same software would have cost us only $1,200.*

- Our workshops are always small: *The class for new managers, for instance, is limited to 12 people. The class on new accounting methods is limited to 15.*

- We need to change some of the procedures in our store to improve its appearance. *For example, we inventory new merchandise at the sales counter—a messy procedure within view of the customer.*

Most of the communication actually takes place in the examples, doesn't it?

Tip

Generally put your key examples toward the front of your presentation. That's when the audience needs them. Too often, though, inexperienced speakers save the examples for the end.

Longer examples

Sometimes a short example just won't do. Either your topic is complex or you want the extra emphasis a longer example gives. Here are three longer examples that speakers handled quite well. As you read them, try to imagine how little communication would have taken place with no examples at all—with only the abstract explanation.

Longer example 1: concurrent engineering

This example is from an engineer giving a presentation at a conference. Here is the abstract explanation of *concurrent engineering:* "designing an item, planning for its production, and planning for its maintenance all at the same time."

Fortunately, the speaker continued with an example:

If you want a new part for a machine, the old way was to design the part, down to the last detail. Then the designers would give their design to the production people who would prepare tools to make it. Once the part was actually in production, the maintenance people would begin doing repairs or preventive maintenance. That was inefficient.

Concurrent engineering plans for *all three stages* at the same time. The production people work with the designers, perhaps pointing out that their tools would have great difficulty rounding out an edge to the designers' specifications. But a slightly different edge would be easy.

The maintenance people might point out that a component that fails every 75 hours is in a place that requires removing 15 bolts. Redesigning would make their job easier.

Having the designers, the producers, and the maintainers work together—concurrently—makes engineering far more efficient. And the product can be much less expensive for its life cycle—perhaps slightly more expensive to design, but cheaper to produce and cheaper to maintain.

The example is effective, isn't it? It would take a minute to present, but I can't think of a shorter way to communicate the concept.

Longer example 2: live-fire vulnerability testing

The next example is from a research analyst. Here's an abstract explanation of *live-fire vulnerability testing:* "finding out what happens to our military weapon systems when enemy weapons hit them."

When I first heard this term, I was mentally searching for a concrete example. Fortunately, the speaker supplied it:

Suppose there's a new tank, and we want to find out what might happen if an enemy missile hits it. People in the military don't want to wait for an actual war to get the answer.

So they get a new tank and actually shoot at it. The tank is fully equipped for battle—loaded with fuel and its own weapons (of course, there aren't any people in it). The missiles they shoot at the tank are similar to the kind the enemy would actually shoot.

The cost is an expensive tank, but the information from the test is often invaluable. Researchers can find out how vulnerable the tank is to the enemy's weapons and possibly make changes to improve it.

The soldiers who operate tanks certainly favor this type of testing.

There you have it: the example. The audience is now ready to hear more about live-fire vulnerability testing. Without the example, they might have preferred being in that tank.

Longer example 3: adapting physical education for children with special needs

This example is from a curriculum coordinator. Here is the abstract explanation of *adapting physical education for children with special needs:* "creating games during physical education for children who are unable to participate in the regular games children play."

Unlike the earlier terms, we can understand that definition fairly easily, but it has little impact. However, the speaker wanted more than intellectual understanding. She wanted us to *feel* the need for her program.

So she was creative—she involved us in a living example. First, she asked for a volunteer to come to the front of the room. Then she asked him to go to the other end of the room, pick up a bean bag, and return. That's typical of a game elementary school students might play.

Next, she asked the volunteer to put on a blindfold. She then asked members of the audience to suggest ways to adapt the game for him. One person suggested holding his arm while he walked. Another suggested handing the bean bags to the "blind person" when he reached the other end of the room. That way, the blindfolded volunteer wouldn't have trouble finding them.

Then the game took place again, the runner blindfolded, the other volunteers participating. It was fun to watch, fun for the blindfolded volunteer, fun for the other volunteers. I suspect it would have been fun for elementary school students participating with a person who was actually blind.

This was a terrific example. By seeing the blindfolded person, we were all able to empathize. We understood what "adapting physical education for children with special needs" meant, and we understood its value, too.

This was also a good way to involve the audience. We'll look further at this in Chapter 8, "Involving the Audience and Using Humor."

Examples may seem simple to the speaker, but they're often crucial for the audience. So consciously look for places for examples: quick ones take almost no time and can be extremely important; longer ones not only help communication but provide extra emphasis, too.

CHAPTER 4

Remembering What To Say

A speaker's greatest fear: going blank with everybody staring. You try to think of something—anything!—but your brain seems to be disconnected.

What can you do?

Nothing. At that point, it's too late. But you can do something to avoid being in that situation (other than declining invitations to speak). You can design a way of remembering the content of your presentation so going blank almost never happens. And if it does—which is unlikely—you'll have something to help you get going again.

That's what this chapter is about. My recommendation is that you try to use visual aids for almost all speaking situations. If they're well designed, they'll help not only your audience —they'll help you, too.

But to see the value of visual aids, let's consider all four common ways of remembering material:

- memorizing
- reading from a complete text
- using notes
- using visual aids as notes

Memorizing

Memorizing is absolutely the worst way to keep track of your material.

People who memorize almost always are preoccupied with the *words* they're saying—not with the *ideas* behind those words (or with the audience). As a result, normal inflection disappears. And, worse, those terrible blank moments become almost inevitable.

I used to memorize the first few sentences of my introduction, just so I'd get off to a good start. But I found that I almost always fumbled those lines. Now when I speak, I'm very familiar with what I want to say, but I'm familiar with the *ideas* rather than with the exact words. When I rehearse, I use similar words to express my ideas, but I don't use exactly the same words each time.

Even top professional actors can have trouble with memorized speeches. Just watch the Academy Awards.

I suggest you avoid memorizing.

Reading from a complete text

Ask audiences what they most hate about presentations, and someone is sure to say, "Having people stand up and just read to me. If that's all they were going to do, I could have read it myself."

Why is reading a presentation hard to do well?

Most of us have suffered through people reading badly. Here are some reasons that happens:

- *The speaker loses normal inflection.* Like people who have memorized their speech, people who read aloud often lose touch with the ideas behind the words. You can easily tell if that happens: listen for pauses. Natural speaking is filled with them; unnatural reading isn't.

- *The text isn't spoken language.* Too often speakers write their speeches in "business-ese"—that difficult gobbledygook that's hard enough for us to read, much less listen to.

- *The speaker is static.* The potted plant will probably move more. There is little movement, little energy, little visual interest behind the lectern.

- *There's no eye contact.* Any eye contact is with the text, not with the audience. Gestures are limited to adjusting eyeglasses.

- *The speaker is scared.* Often speakers decide to read their speech because they're afraid to try anything else. They know that reading will fail, but at least it will fail with a small "f" rather than a capital one.

But reading isn't always bad. (Just almost always.) Sometimes speakers simply have to read: they're announcing a precise policy statement; timing a short presentation down to the second; talking to speakers of English as a second language and avoiding colloquialisms; or attempting the kind of eloquence that rarely happens without the exact words. If reading is necessary, here are some suggestions.

Some suggestions for reading aloud

Good readers can overcome the problems I just mentioned. Here's what I suggest if you must read your presentation:

- *Inflection.* To make your words sound natural, rehearse often. Check yourself for pauses. Ask yourself if your words sound the way you'd say them.

- *Spoken language.* You can also improve your inflection by choosing words you might actually say—rather than using "business-ese." In fact, top speech writers work to put colloquialisms in the text: "Okay, let's push that idea a little farther and see what we come up with" or "I think this new approach will be easier once we get the hang of it."

- *Movement.* Plan for key gestures: pointing in the general direction of the city you're mentioning, showing how big or how small something is, shrugging at the right time, raising an eyebrow. Some speakers put cues for these gestures in their text.

- *Eye contact.* If your text is "user friendly," you have a better chance of looking at your audience. So you need a good layout for your text. You don't want bunches of long paragraphs, or you'll lose your place every time you look up. Instead, try starting a new paragraph after every sentence or two. And avoid all capital letters, which are hard to read. Instead, use upper and lower case in a larger type size.

Tip

The audience can get distracted by watching speakers turn the pages of their text. Instead, use unstapled pages and simply slide the page you've finished to the side.

A good example

Picture a graduation speech: dull, boring, the audience half asleep. It doesn't have to be that way. I recently saw a graduation speaker do a terrific job, and he read the entire address aloud, holding the audience spellbound.

How did he succeed? First, his text was extraordinarily witty and in spoken English. Second, he was energetic—gesturing constantly, varying the loudness of his voice, even singing the French national anthem in English (to illustrate a point). Third, his eyes were piercing. We had the feeling he was looking at us constantly. Finally, he wasn't a bit scared of us. The auditorium was his, and he ruled it totally.

But don't read your presentation

As far as I'm concerned, reading well requires me to work too hard at the wrong time: when I'm actually giving my presentation. I have to be constantly aware of my gestures, inflection, eye contact—things that come naturally when I'm just talking. That's why bad speech readers are common and good ones are rare. Talking normally is usually much more effective.

Tip

If you expect to be nervous, try reading the first words of your presentation—especially a funny or moving quotation. That way, you'll take lots of pressure off the most nervous time of all: the beginning. But then move quickly away from reading—preferably to visual aids. In fact, the quotation can be your first visual aid.

Let's turn now to what used to be the most common technique for remembering material.

Using notes

Using notes is normally better than reading: speakers can have normal inflection and make eye contact more easily.

However, if the notes are on a lectern, speakers probably won't move very far from it. And if the notes are in their hands, they won't gesture much. Also, there's that constant looking down at the notes. That doesn't happen when speakers use visual aids correctly—when the speaker looks at the visual aid, the audience should be looking there, too.

Some suggestions for using notes

If you must use notes, here are some suggestions:

- *Use note cards.* They're small and easy to handle.
- *Don't put much on them.* What would you use as headings or subheadings if you had written your entire presentation? That's what should go on your note cards—not lots of text. Too many notes get in the way of good eye contact.
- *Leave your notes on the lectern or table and move away from them occasionally.* During parts of your presentation when you're especially confident, move around to gain rapport with your audience. You can rehearse when to do this.

Tip

Be sure to put these items on your note cards: quotations, statistics, and lists. We've all seen speakers have trouble thinking of that last item in the list they're giving.

A good example

I once saw a woman who was excellent at speaking from notes. She'd start by glancing at them, then leave them on the lectern and walk to the center of the stage where she'd simply talk. When she finished her point, she'd walk quietly back to the lectern, glance at her notes, and return to the center of the stage.

You might think she looked strange walking quietly back to the lectern. On the contrary, we were much more aware of her walking back toward us. The feeling was that she, like most human beings, needed some reminders of what to say. But she wasn't going to let that interfere with talking to us.

Using visual aids as notes

Now for the technique I've been referring to all along: visual aids. Today they are by far the most common way to remember what to say. Let's consider an example.

I once saw a person explain his tasks as a desk editor for a major news magazine. He spoke for about ten minutes using four visual aids. This was his first one (his blueprint):

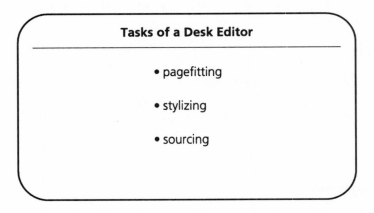

He briefly defined *desk editor* and the three terms on his visual aid. Then he turned to his next visual aid:

Pagefitting

- placing the text on the page

- changing the text to fit

During this section, he gave us several examples for each point.

When he had said all he had to, he didn't need to worry what was next. He simply turned to his next visual aid:

Stylizing

- external

- internal

Again, this was all he needed to remember what to say. The examples he'd rehearsed came right to mind. By now, you can imagine what his final visual aid looked like:

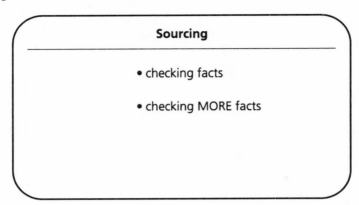

These may seem so simple and obvious, but that's the point: the simple visual aids served as the headings and subheadings he might have used if he'd written down his entire presentation. When he spoke, he'd come to a "heading," say what he wanted to, and move on. If he forgot something he'd intended to say, that was all right. The audience would never know.

This technique of using visual aids as notes has these important advantages:

- *You don't have to worry about what you're going to say next.* This is a *significant* advantage. Your next visual aid has your next major idea. Just turn to it when you're ready. That way, your mind isn't constantly cluttered by the fear that you may forget what's next.

- *You can move about the room.* Inexperienced speakers don't want to move, but movement helps you relax and adds energy to your presentation. We'll talk more about this in Chapter 11, "Using Effective Techniques of Delivery."

- *You can have good eye contact with your audience.* You can look at your audience all the time—except when you're looking briefly at your visual aid. But that's okay; the audience will look at your visual aid then, too.

- *Your audience feels comfortable knowing you're on your planned track.* Well-designed visual aids show that you have a plan and are following it.

Tip

Your visual aids don't need to be only word charts: diagrams, pictures, and graphs all give structure to a section of your presentation and serve well as notes for you. Chapter 7, "Further Suggestions for Designing Visual Aids," can give you some ideas.

Four other chapters discuss visual aids:

- Chapter 5, "Choosing Visual Aids," talks about the different kinds available—from overhead transparencies to flip charts—and the advantages and disadvantages of each.

- Chapter 6, "Designing Visual Aids," then suggests good ways to prepare your visuals so they'll look good and be effective.

- Chapter 7, "Further Suggestions for Designing Visual Aids," continues Chapter 6.

- Chapter 12, "Presenting Visual Aids," gives some do's and don't's for handling visual aids during your presentation.

CHAPTER 5

Choosing Visual Aids

Some older books on speechmaking cautioned against visual aids: "Don't rely on them too much," they'd say. "Don't use them as crutches." But the standard for those books was the immobile speaker standing behind a lectern, declaiming to the audience for 30 minutes.

Things have changed. Today, almost all business speakers use visual aids. One reason is technology: many of us can produce them rather easily at work. Another is that we've learned their advantages—for us (as speakers) and for our audiences:

- They keep *us* aware of where we are in our presentation.
- They keep *the audience* aware of where we are in our presentation.
- They visually reinforce our words: the audience can see an idea and hear about it at the same time.
- They emphasize important ideas.

Once you get used to speaking with visual aids, you'll rarely want to speak without them. The advantages are too important.

This chapter will help you choose the best type of visual aid for your presentation. We'll consider the advantages and disadvantages of these (the most common types people use):

- overhead transparencies
- 35-millimeter slides
- flip charts
- blackboards and whiteboards
- computer presentations
- objects and models
- imaginary visual aids
- handouts as visual aids

Overhead transparencies

This is my favorite type of visual aid for business presentations—and the favorite of most speakers. It requires only an overhead projector, a screen (or even a wall), and transparencies. Virtually all conference rooms are equipped for them.

Advantages of overhead transparencies

- *Transparencies are easy to make.* Simply prepare your visual aid on paper and then copy it on a copier. But instead of copying onto blank paper, copy onto a transparency. This way you can make transparencies quickly, revise them quickly, and revise them often. You can also make them yourself, without waiting for a professional staff to produce them for you.
- *They're cheap.* They cost only pennies a copy.
- *They're portable.* For most presentations, you can easily fit your transparencies in your briefcase with room to spare.

This is no small matter if you travel often. If you use cardboard frames around your transparencies, you increase the bulk somewhat, but they're still quite portable.

- *They let you be flexible.* You can rearrange your presentation on the fly—with the audience staring at you—to meet new needs. For example, occasionally someone needs an answer now for something you planned to cover later. No problem: just reach for the appropriate transparency and press ahead.

- *You can write on them.* Sometimes you don't want the audience to see a static visual aid (like a ready-made equation); instead, you want to create it, step by step, as the audience watches. With a transparency, you can do that easily. Some pens are designed for that purpose. An added advantage is that you can easily erase your writing later with a damp paper towel (if you're like me, however, you may walk around for a day or so with red or green or blue fingers).

- *You can see what's next.* Since most speakers handle their own transparencies, they can glance at the label on the next one and see what the next topic is. That's a really important advantage; otherwise, part of your mind is constantly trying to remember what's next. Sneak glances are no problem with overheads.

- *They can look extremely professional.* Color printers and copiers can enhance your message by drawing attention to key features and providing a visually interesting (yet still unobtrusive) background. You can also reproduce colorful photographs on a transparency and have good resolution.

- *They can be informal if necessary.* For an impromptu meeting with colleagues, you can simply hand print or hand draw your transparencies.

Disadvantages of using overhead transparencies

Alas, the world isn't perfect. Here are some of the disadvantages of using overhead transparencies:

- *The projector may not be very good.* Because overhead transparencies are the most popular visual aid, the equipment takes a beating. You'll often find projectors that don't focus well, have dim bulbs, or have no bulbs at all. That's why I carry my own projector for local presentations. When I travel, I insist in advance on a good projector. Then, when I arrive at the place for my presentation, I go immediately to the projector and try it out. If it's not good, I try to get another one.

- *The bulb can burn out.* Many of today's projectors have a spare bulb built in, but sometimes the spare is burned out, too. Presenters who have a bulb burn out switch to the spare and go on. They usually forget to tell visual aids people that the overhead is now down to only one good bulb. If that one burns out, you're left with none.

Most speakers consider the overhead transparency their first choice of visual aid unless they have a good reason not to use it. That's what I suggest, too. But there are many, many good reasons to use other visual aids, either separately or in conjunction with overhead transparencies.

35-millimeter slides

Advantages of 35-millimeter slides

Thirty-five-millimeter slides can produce wonderfully professional presentations. You can show colorful word charts and colorful pictures, too. Audiences love real pictures—of things, people, even themselves. NASA has a speaker's bureau that uses slides, and you can imagine the beautiful photographs of space walks, the Earth, rockets taking off.

If you want a thoroughly professional look with high resolution, slides are the way to go. They're also especially useful for standard presentations that many people give. For example, a woman I knew owned a business selling light fixtures. She made 35-millimeter slides of the fixtures her company had actually installed, and then she built her standard sales presentation around those slides. All her marketing reps used copies of that standard presentation.

Once slides are in the tray (right side up, facing the right direction) they stay right—and in the proper sequence. There is little fuss to handling them other than clicking a button to change to the next one.

Disadvantages of 35-millimeter slides

However, there are some disadvantages to using 35-millimeter slides. The most significant is that making them normally requires professional help. That means they're usually expensive and hard to revise. They also usually take longer to make. And they're virtually impossible to rearrange during a presentation.

Also, 35-millimeter slide projectors aren't nearly as common in conference rooms as overhead projectors, so you may have to carry your own. And if there is one there, it may have problems with jamming, focus, or burned out bulbs—and no spare projector.

Also important, the lights normally should be much dimmer in the room for a slide show than for transparencies. That's very restful for your audience. *Very* restful.

Tip

Consider using 35-millimeter slides if you want a thoroughly professional look, have the time to prepare them, don't expect to make many changes, have some exceptional pictures, and won't have the lights off very long.

Flip charts

Flip charts are very large tablets of paper on an easel.

Advantages of flip charts

Flip charts are useful for creating a visual aid "before your audience's very eyes." That is, you can start with a blank page, ask for the audience's ideas, and record those ideas on the flip chart. The effect is that you have your "sleeves rolled up" and are ready to get to work.

You can begin a presentation by making a list of what people in the audience want to talk about—they enjoy having you discuss and then check off their ideas. You can even have a volunteer come to the front of the room and record the audience's ideas.

Flip charts are also cheap, easy to make (even in advance), and easy to revise. You can tear off the pages and tape them around the room for later reference. Some people, particularly in informal presentations, put their blueprint on a flip chart and leave it in view. Then they use overhead transparencies or some other means for the main part of their presentation. When they're ready to move to another section of their presentation, they simply walk over to the flip chart, point out the next item on the blueprint, and move on.

Flip charts are also easy ways to use color—simply have a handful of colorful pens and you can put your artistry to work. Some speakers like to make all headings one color and all text or bullets another.

Tip

Use a flip chart for a "sleeves up" approach when you have a small audience and want to record the audience's ideas during your presentation. Or use it if you want to keep the pages on display around the room.

Disadvantages of flip charts

The main disadvantage is that flip charts won't work for large audiences—those sitting in the back of the room may not be able to see them. If your audience is small, you have no problem.

They're also cumbersome to carry around, and they're hard to save or reuse. The pages sometimes crinkle—noisily—when you fold them back. And they require an easel (not always available), which makes a nice object for people like me to back into or trip over.

Flip charts also usually look homemade and depend on relatively neat handwriting (legible, at least).

If you're writing on a flip chart during your presentation, you may have trouble writing neatly and talking at the same time. And even if you're a good speller, your hurry to put words on the page in front of your audience may cause a handwritten "typo." (Or "thinko," as I've heard them called.) Then that misspelled word will stay in front of your audience's eyes, constantly distracting them. Also, the writing from colored pens can bleed through to several sheets beneath.

Tip

When using a flip chart, leave one or two blank pages after every page you write on. That will avoid showing the bleed-through that occurs. You may also want to put tabs on the pages you plan to use. That way you won't have to turn a number of blank pages to get to the one you want.

Blackboards and whiteboards

Blackboards and whiteboards have essentially the same advantages and disadvantages as flip charts (except you can't pack a blackboard and take it with you, and you can't page back and forth). I especially like whiteboards because they lighten the room and show color well.

The biggest caution for whiteboards is the pen: some pens are intended for whiteboards and erase very easily; some don't erase—ever—and give memorable indigestion to visual aids people. In fact, visual aids people try not to allow permanent marking pens in the same room with a whiteboard.

If you're using a whiteboard, check the pens in advance. Otherwise, you may be embarrassed when you try to erase the board and find—behold!—the words just stay there, giving a whole new meaning to "my words will live forever." You'll want to slink quietly off.

Computer presentations

Computer presentations are becoming ever more popular. Essentially, they project either onto a large monitor or through an overhead projector onto a screen.

Advantages of computer presentations

If you're showing what a computer program can do, computer presentations are indispensable. And they're also useful replacements for overhead projectors, 35-millimeter slides, and even flip charts. They show color extremely well, are relatively easy to create and modify (once you learn how), and can look highly professional. They are superb for animating your visual aids—for example, adding bullets step by step to a chart.

They can also be wonderful for doing "what if" presentations—perhaps using a graph, changing some values, and having a formula calculate a new result. For example, in a presentation on sales projections, you could change the projected sales figures in front of your audience's eyes, and everyone could then see the effect on your company's bottom line.

You can easily add graphics called "clip art"—predrawn figures you simply cut and paste into your computer presentation. Chapter 7, "Further Suggestions for Designing Visual Aids," has an example of clip art.

Computer presentations are handy for standard (perhaps daily or weekly) presentations when data changes but format doesn't. The effort to prepare the computer presentation the first time is repaid ever after because updating the data—and, hence, the presentation—becomes simple.

Automated computer presentations are also handy at conventions when roving audiences stop in randomly. The computer show draws the crowd and keeps going longer than most human beings can. Also, for such shows, you can design participation: the user answers a question or pushes a button for some action.

Finally, you can have a wealth of backup material in your computer and call on it instantly if the need arises.

Tip

Use a computer presentation to show how a computer works or to put on a continuous display for an audience that comes and goes. Also use one if you have a relatively small audience and want to show snazzy graphics and color.

Disadvantages of computer presentations

Computer presentations depend on technology, so there's more to go wrong. And if something goes wrong with the equipment, you'll be pretty well stuck. Also, the technology is still evolving, so the projected image can be dim or fuzzy or visible only to a narrow segment of the room.

The equipment—a computer (possibly portable) and a projection unit—is expensive, fragile, and awkward to carry (if it isn't set up permanently where you'll give your presentation). The equipment can also be difficult to set up and adjust.

The biggest disadvantage of computer presentations—especially the kind that aren't automated and depend on an operator—is the potential for real-time glitches not possible in other media. If the operator presses the wrong key, enters the wrong data, or (worst of all) accidentally erases the file, the result can be an unrecoverable disaster. Nevertheless, computer presentations can be terrific and are increasingly popular.

Objects and models

Sometimes there's simply no replacement for an object itself. If it's small (like a new kind of light bulb), there's no reason to have pictures or diagrams of it. Just show the light bulb.

If it's large, like a new building—or tiny, like an atom—a model can be the highlight of your presentation: professional looking and interesting.

Tip

Don't pass an object around the audience. You'll be creating your own greatest distraction: people looking at the object simply won't be paying attention to you. Just hold the object up or walk quickly around the room with it. If it's not fragile or expensive, consider leaving it on display for a few minutes after your presentation.

Imaginary visual aids

Yes, some of the most creative visual aids I've seen in presentations haven't even existed. Here are some examples of imaginary visual aids:

- One person was showing the distance someone could broad jump. So she made the stage an imaginary place for the event, started at one edge, and walked the distance for the high school record. She talked about that awhile, then moved a little farther to show the collegiate record. And so on.

- Another person turned the stage into a boat. He showed us where starboard was, the stern, and so on. He used the stage as his reference throughout his presentation.

- Another made the stage an airport, showing which directions the planes would take off and land, where the gates were, and where the control tower was. She then used this to illustrate the various traffic patterns the planes would fly, depending on the direction the wind was blowing.

Such creativity isn't too informal: these speakers were in relatively formal situations, and their creativity added liveliness and interest.

Handouts as visual aids

An all too common visual aid is the handout. Inexperienced speakers often pass one out even when there's no immediate need. The result is a major distraction. Simply look around the room, and you'll often see people leafing through the handout rather than looking at the speaker. In fact, Jim Casimir, a top executive in the IRS, says, "A good rule is *never* use a handout during your presentation."

If you *must* use a handout during your presentation, I suggest these steps:

- Pass out the handout yourself, counting out the right number for each row. That's much faster than dropping off a bunch at one side of the room and waiting for copies to get to everyone. In a large room or auditorium, ask several members of the audience to help you. You can increase their efficiency by giving them specific instructions: "Please take the left side and count out the correct number for each row. Thanks."

- Go through the handout as a group, pointing out what, specifically, the audience needs to look at. That way, everyone's attention will be focused.

- When you're through, ask members to set the handout aside. I say something like this: "Okay, we're through with the handout, so if you'd set it aside, we'll move on." Audiences rarely seem to feel dictated to. They simply set the handout aside and look up.

Some organizations are used to passing out a paper copy of their visual aids to the audience. If possible, do that at the end of the presentation. That way the audience isn't distracted.

We've covered many kinds of visual aids, each with advantages and disadvantages. The next chapter offers suggestions on how to design them.

CHAPTER 6

Designing Visual Aids

This chapter deals with the fine points of designing visual aids that look good and help you and your audience. I'll concentrate on the overhead transparency, because it's the most common visual aid and because the fundamentals for it apply to other visual aids, too.

Some organizations have standard formats for their visual aids. But lots of organizations don't. In fact, you may be your own typist, graphic artist, typographer, and photocopier. Many of us are. If so, this chapter will help you design your own visual aids well. If someone else does your typing and designing, this will help you tell them what you want.

Margaret Raab, in *The Presentation Design Book,* says that "good graphic design is invisible." So true. For most occasions, we don't want the audience to gasp with pleasure at eye-catching visual aids. Instead, we want them to concentrate on the content—to pay attention to *what* we're saying rather than to *how* we're saying it.

Keith Thompson, in *PC Publishing and Presentations,* expresses a similar idea: "Your slides should reinforce your message, not overshadow it. To put it another way: the speaker needs to remain the center of interest, not the slides."

49

Of course, there are exceptions. It's fine to show off a little. But you'll usually want your visual aids to have straightforward efficiency.

Here are suggestions for designing efficient visual aids:

- Use a title transparency.
- Put the words near the top.
- Don't use too many words for each idea.
- Don't put several ideas on the same transparency.
- Use upper and lower case.
- Use large letters.
- Use a sans serif font.
- Use a single orientation: all landscape or all portrait.
- Add some color.
- Be careful with borders.
- Consider using frames to hold your transparencies.
- Label your transparencies.

Let's look at each of these more closely.

Use a title transparency

Your first transparency should have at least the title of your presentation and your name on it.

Tip

If the title of your presentation has unfamiliar terms in it, you may want to leave the title transparency on while you define those terms. I've seen people spend two or three useful minutes on the title transparency before moving to the next one.

Some speakers decide to replace their name on their title transparency with the name of the person or organization they're speaking to. That's normally not a good idea: members of the audience know who *they* are, but they don't necessarily know who *you* are. They'll be happy to see your name so they'll know how to refer to you during the presentation.

By the way, I suggest you use the name you want people to call you: "Ed Bailey" instead of "Edward Bailey" or "Professor Joan Hiller" instead of "Professor J. Hiller."

Tip

You might want to have a test transparency—all mine says is "test transparency   test transparency   test transparency. . . ." That way, when you're testing the overhead projector and some members of the audience are in the room, they won't see one of your actual transparencies.

Be sure to use typical type sizes on your test transparency, including your smallest type size.

Put the words near the top

We're all trained from childhood about the value of centering things, so naturally we want to center the words (or diagrams, or pictures) like on the overhead transparency on the top of the next page.

It looks nice there, and so will the paper copy you use to run your transparency. But keep in mind that audiences can't always see all the screen—those in the back may see only the top. So I always work from the top of my transparency, because the top of the screen is visible to everybody in the room. That is, I center my text from left to right but not from top to bottom. In other words, not like this:

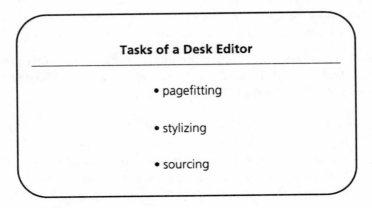

I think of the top of the screen as "golden" space I want to make the best use of. Here's how my transparency would actually look:

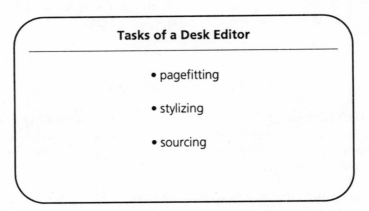

Putting the words on the top doesn't make the transparency appear distorted or off-balance, does it?

You may think you can just push the transparency up on the projector during your presentation. But that's just one more thing to think about, and it may leave an empty space at the bottom of the screen.

Don't use too many words for each idea

Perhaps the most common mistake is putting too much material on visual aids. If you put everything you have to say on your transparency, you'll end up reading it aloud, turning to your next one, reading that, and so forth—leaving nothing else for you to say. Audiences don't appreciate hearing speakers read most of their presentation.

Here is a transparency with wall-to-wall words:

What Headhunters Do

- The first task headhunters do is find openings in other companies. These openings should normally be for relatively senior people who have special skills or academic qualifications.
- The second task is to find prospective employees to fill those job openings. In addition to good paper qualifications, they must be effective. Otherwise, the companies won't use our headhunters.
- Finally, headhunters match employees and openings by holding interviews with everyone involved. These interviews make sure the hiring company and the prospective employee will be happy.

Instead, choose a few key words to serve as reminders. As I mentioned earlier, these reminders are like the headings and subheadings you might use if you'd written your entire presentation.

Here's a revision, this time with only brief verb phrases as reminders:

What Headhunters Do

- find job openings

- find prospective employees

- match employees and openings

You've now left something to say, so you won't just be reading aloud. This isn't a small point. Most presentations by inexperienced speakers have far too many words on the transparencies. That virtually guarantees problems: the audience won't want to read the transparencies because they look cluttered. And if the speaker does nothing but read transparencies, the audience gets upset.

That's not to say that transparencies must have only simple phrases on them. Short sentences are okay, especially for questions and brief quotations. Just work for less rather than more—let phrases be the norm and full sentences be the exception.

Tip

Some organizations encourage speakers to leave a paper copy of their visual aids with the audience. As a result, speakers tend to make the visual aids self-explanatory so the paper copies will stand alone. The actual visual aids then become much too wordy. Instead, make your visual aids the best you can for your presentation, then annotate the paper copies to help them stand alone.

Don't put several ideas on the same transparency

Another common mistake is putting too many ideas on the same transparency. Here's one from a presentation on expert systems:

Expert Systems

Mechanical expert system
> <u>Symptoms</u>: the engine won't start, the lights are off, the radio won't play
> <u>Solution</u>: the battery needs to be recharged or replaced

Medical expert system
> <u>Symptoms</u>: runny nose, sore throat, coughing
> <u>Solution</u>: the patient has a cold

Cluttered. Instead, try using two transparencies:

Mechanical expert system

Symptoms
- the engine won't start
- the lights are off
- the radio won't play

Solution
- the battery needs to be recharged or replaced

Medical expert system

Symptoms
- runny nose
- sore throat
- coughing

Solution
- the patient has a cold

By using two transparencies, you can also improve the lay-out—adding white space and bullets.

Use upper and lower case

Before computers, which can easily vary type size, speakers often used all upper case (capital letters) on transparencies. That was the only way a typewriter could make words large enough. Today, however, researchers believe that a page entirely in upper case type is hard to read:

THE FIVE C'S OF CREDIT

- CHARACTER
- CAPACITY TO REPAY
- CAPITAL TO CUSHION AGAINST LOSSES
- COLLATERAL
- CONDITION OF THE ECONOMY

It certainly is old fashioned and hard to read. Instead of using all upper case to get larger letters, just use upper and lower case in a larger type size:

The Five C's of Credit

- Character
- Capacity to repay
- Capital to cushion against losses
- Collateral
- Condition of the economy

Use large letters

Looks can be deceiving. What looks plenty large on the transparency you're holding in your hand can look awfully small once you project it on the screen. Too often speakers wait until the actual presentation before testing their transparencies—then look with shock at the small letters. Inevitably, they mutter something like, "I hope you all can read this." That doesn't make for a strong start, and it throws the speaker off guard, too.

There isn't any right type size to use because:

- projectors have different focal lengths
- the size of the projection on the screen depends on how far the projector is from it
- in a large room people may be farther from the screen than in a small room

Even though these variables exist, I rarely use a type size smaller than 18 points. I normally use 20 points or larger. In case you're not familiar with point sizes, here are samples:

This is 18-point type.

This is 20-point type.

Use a sans serif font

Terms like *points* and *serifs* were alien to most of us a few years ago, but the computer revolution for word processing has made the terms more common and given us the ability to put them into practice.

Serifs are those little lines that hang down from the crossbar of a *T*, stick out from the sides of an *H*, etc. But some typefaces (called *sans serif*) don't have those lines. Here are examples of both kinds:

Serif: THEIR their

Sans serif: THEIR their

The standard for visual aids is a sans serif font. This is the opposite of documents, which use serifs for text (like this book), reserving sans serif fonts for some headings.

For a transparency, though, sans serif fonts project a cleaner, less cluttered image on the screen. I use them almost exclusively.

Use a single orientation

There are two basic orientations for your slides: landscape and portrait. Landscape means the transparency is wider than it is tall:

Landscape Orientation

- landscape

- landscape

- landscape

Portrait Orientation

- portrait

- portrait

- portrait

The standard for presentations is landscape, but either orientation is acceptable:

- Portrait has space at the bottom of the transparency that projects too low on the screen. That's fine. Just keep your content on the top two-thirds of the transparency.

- Landscape is so wide that a line of text going all the way across the transparency would be too long. Be sure to use generous margins.

Basically, landscape and portrait both give you the coverage on the screen you need. The problem is they give you *more* than you need. Landscape gives you too much width; portrait gives you too much length. But if you place your words well, either orientation will do.

Tip

> *Try not to have some transparencies in landscape and some in portrait. Your presentation will look inconsistent.*

Add some color

Color can make your presentation look professional and, used carefully, help reinforce your message.

Color is easy to add with transparencies:

- *Use a colored transparency.* Common background colors are yellow, blue, green, and pinkish red. I prefer yellow: yellow brightens the room and provides nice contrast for black lettering. Green and blue seem too restful for me—and I don't want my audience to leave the room well rested. Pinkish red doesn't seem to have any good use. Clear transparencies are fine (my second choice to yellow), but they're ordinary and tend to draw attention to any flaws on the screen or on the lens of the overhead projector.

Tip

Use one color (perhaps blue) for your title transparency, blue-print, and section headers. Use another color (perhaps yellow) for the rest.

- *Use a colored pen.* Writing on a transparency with a colored pen (I like red) really puts a splash of color on the screen.
- *Use a color printer or copier.* If you have access to a color printer or copier, you'll have no trouble making colorful transparencies. Be sure to have consistent colors throughout for background and text. Or at least have consistent colors for sections of your presentations. You may even want all title transparencies to have one color combination; the other transparencies would then have a different color combination.

One caution: used inappropriately, colors distract from a message rather than reinforce it. Anytime members of an audience gasp with pleasure at a creative, colorful transparency, they've probably lost track of the speaker's reason for the transparency: the message on it.

Tip

Use bright colors sparingly, mainly to call attention to something—such as the arrows on a flow diagram or the bullets on a chart.

Be careful with borders

Borders are those nice looking lines that form edging around a page:

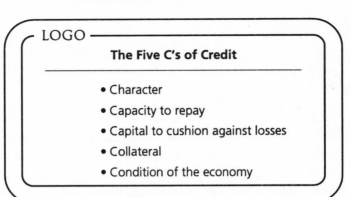

The problem is that what looks nice on paper and nice on a transparency in your hands looks bad when projected: most conference rooms and screens aren't set up so everyone in the room can see the entire transparency. So the audience doesn't see a nice border; the audience sees only *part* of a nice border. The rest is cut off. I suggest not adding a border to the transparency itself.

Consider using frames

Most speakers tape each transparency to a cardboard or plastic frame. That's often a good idea:

- The frame provides an opaque edge that allows only the transparency itself to project onto the screen. Otherwise, there would be a glaring edge because the face of the overhead projector is larger than a transparency.

- The frame makes handling transparencies easier.

- It takes care of the static electricity problem. Freshly prepared transparencies have an amazing amount of static electricity. That causes them to cling, turning neat stacks into messes. A frame solves the problem.

- It helps you put the transparency on straight.

If you don't use frames, put tape on the face of the projector to cut out the glare. I put a test transparency on the projector, make sure it's projecting straight on the screen, and put tape all around the outside of the transparency. In effect, I create on the projector a rectangular frame made of tape.

Tip

If you do use tape, don't use MASKING *tape. Use* DRAFTING *tape which isn't very sticky. Masking tape can ruin the face of the projector: the top of the tape peels off later, but the glue tends to stay put.*

If you have a lot of transparencies and use them frequently, you probably want to avoid frames. They're quite bulky. I can easily fit all the transparencies for a two-day seminar in my briefcase. But if they had frames, I'd need several briefcases.

Tip

You can write brief notes on the frame. Some people like to write their transition statements on the frame so they can read those words as they place the transparency on the projector. Others like to list a few points they want to be sure to cover. Don't put too many notes on the frame, though, or you'll have more eye contact there than with your audience.

Label your transparencies

When you're giving your presentation and look down at your stack of transparencies, you can't see what's on the next one. Because they're "transparent," you can see through several at once.

I suggest you write a brief title on the frame for each one. I don't number my transparencies because I'm always adding another, taking one out, or rearranging. Simple titles are effective.

If you don't have frames, put a label (the kind for file folders) on the bottom corner (which often doesn't project on the screen):

The Five C's of Credit

- Character
- Capacity to repay
- Capital to cushion against losses
- Collateral
- Condition of the economy

5 C's of credit

The next chapter gives you further suggestions for designing visual aids.

CHAPTER 7

Further Suggestions for Designing Visual Aids

This chapter continues Chapter 6, "Designing Visual Aids." Here you'll see a number of "before" and "after" versions, each illustrating a common problem. I'll concentrate on the overhead transparency, but the principles work with 35-milli-meter slides and many other types of visual aids.

These are the suggestions I'll cover:

- Be consistent with your design.
- Place logos effectively.
- Use varying type styles and sizes.
- Try replacing words with an image.
- Consider using graphs.
- Use only relevant clip art.

Now let's look at examples of each.

Be consistent with your design

One common mistake people make is designing good *individual* visual aids but not a good *set* of visual aids. Each transparency may look nice, but together they don't have a uniform appearance.

The next three transparencies are quite inconsistent:

<u>PAGEFITTING</u>

PLACING THE TEXT ON THE PAGE

CHANGING THE TEXT TO FIT

STYLIZING

external

internal

SOURCING

- checking facts

- checking *more* facts

Now let's use a consistent design:

Pagefitting

- placing the text on the page

- changing the text to fit

Stylizing

- external

- internal

Sourcing

- checking facts

- checking *more* facts

A consistent design looks good to the audience. So try to keep the titles in the consistent places on each transparency; the spacing consistent; and the type faces consistent for titles, headings, and body text. There's a benefit to you, too: if you (and your company) have a consistent design, you can interchange transparencies for different presentations, and for different presenters, too.

Place logos effectively

Chapter 6 pointed out the value of the top part of a transparency—the part virtually everybody in the audience can see. Yet many companies have, as part of their formal style guide, a requirement to put the company's logo or name in the top 20% of the transparency. That doesn't leave enough room for the actual content.

I suggest putting the company's logo or name only on the cover transparency where it can appear in a large size with good emphasis. If you really must have it on every transparency, put it at the bottom or on the same line as the transparency's title. That way, your content will be clearly visible. Notice all the wasted space when the company's name is at the top:

EDITORIAL CONSULTANTS, INC.

Tasks of a Desk Editor

- pagefitting

- stylizing

- sourcing

This looks fine, doesn't it? Unfortunately, many in the audience won't see what we see on the page. The ones in the back will probably see only the top part. So put the company's logo only on the cover transparency—or at the bottom of each transparency, like this:

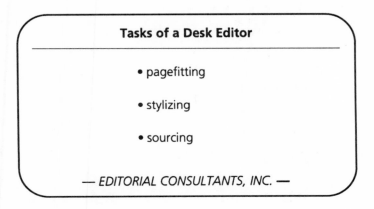

True, some people won't see the company's name at the bottom, but at least the main reason for the transparency—its content—is more visible.

An alternative to having the company's *name* at the top is to put a small logo in a top corner on the same line as the title of the transparency. But be careful not to use too much of that good space at the top.

Use varying type styles and sizes

Computers increasingly give us easy access to varying type styles (such as bold and italic) and sizes. Here's a transparency that looks decent, but it could be better:

```
┌─────────────────────────────────────────────┐
│           Mechanical expert system            │
│  ───────────────────────────────────────────  │
│                                               │
│   Symptoms                                    │
│        • the engine won't start               │
│        • the lights are off                   │
│        • the radio won't play                 │
│                                               │
│   Solution                                    │
│        • the battery needs to be recharged    │
│          or replaced                          │
└─────────────────────────────────────────────┘
```

Now notice how much better it can look:

```
┌─────────────────────────────────────────────┐
│         **Mechanical expert system**          │
│  ───────────────────────────────────────────  │
│                                               │
│   **Symptoms**                                │
│        • the engine won't start               │
│        • the lights are off                   │
│        • the radio won't play                 │
│                                               │
│   **Solution**                                │
│        • the battery needs to be recharged    │
│          or replaced                          │
└─────────────────────────────────────────────┘
```

The bold typeface and slightly larger titles help audiences see the transparency's organization.

Of course, there can easily be too much of a good thing. Be careful to avoid a distracting mixture of type styles and sizes.

Try replacing words with an image

Bonnie Franklin, an expert communicator, says that the key to good visual aids is "to get the right image for the idea." That means looking beyond using only words on a visual aid and considering a diagram, drawing, map, or other image that may communicate more efficiently.

Here's a transparency that tries to describe a computer network using only words:

Computer Networks

Computer networks consist of the following:

- fileserver

- work stations

- printers

These components are connected with cables.

Notice how much more effective a diagram is:

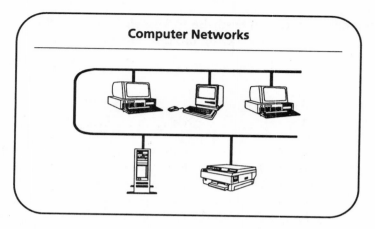

A drawing can be effective, too. If you're talking about something complicated like the effects of lift, drag, thrust, and gravity on an airplane, this line drawing is going to be much more effective than a word chart could ever be:

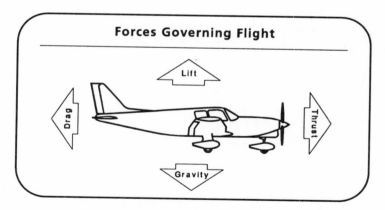

An actual picture of an airplane wouldn't work as well because it would have too many distracting features. But a picture would be terrific for showing a building you're planning to move into, a television you're marketing, or the mountain you've just climbed.

Let's consider another example using something other than words. If you want to tell people where your regional headquarters are, you could use a word chart:

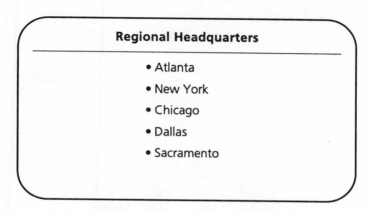

But notice how much more effective a map is:

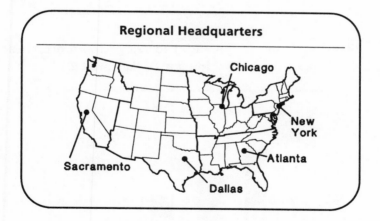

Finding the right image isn't always easy, but once you dicover it, it can get across instantly what a word chart might never convey effectively at all.

Consider using graphs

A graph is another kind of image that can replace words, and we're all familiar with their value. The question I'll deal with is how to present graphs so they're easier to read. Today's technology can take us well beyond yesterday's traditional presentation:

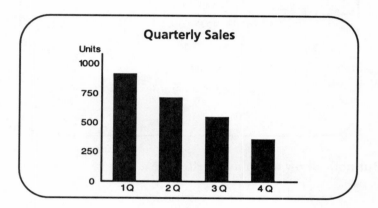

Now let's make this graph easier to read—and more attractive at the same time—by adding a background grid:

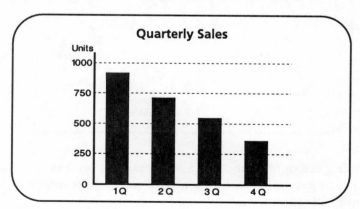

We can also make the graph more attractive by making the bars three dimensional, like this:

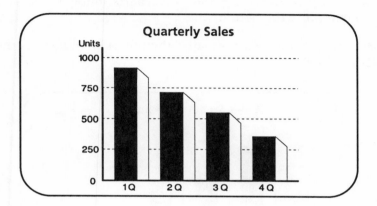

Some people object to three-dimensional bar charts because the third dimension shows an object with volume, which might be a false visualization of a number. Actually, though, the *second* dimension of a bar chart doesn't do much good, either, does it? Usually only the first dimension—a straight line—is all that's necessary to show a numerical value.

So the third dimension may not help your audience understand the numbers better, but it may help your bar chart look more contemporary.

You might enjoy reading Edward Tufte's books on displaying information: *The Visual Display of Quantitative Information* and *Envisioning Information.* Tufte takes the position that displays should be as efficient as possible—with no decoration. My position is more moderate, yet I find his books immensely interesting and informative.

Another tip for graphs is to have an informative title. In Chapter 2, I suggested having the main point of your presentation—the "bottom line"—near the beginning. You can use the same principle for graphs and charts simply by using a title that states the bottom line.

See how much more informative the title is when it states the bottom line?

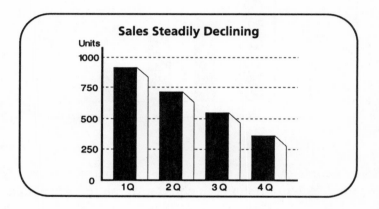

Use only relevant clip art

Clip art is prefabricated line drawings. People can buy hundreds or even thousands of clip art images and then scatter them throughout their transparencies. Audiences naturally turn their attention first to the clip art. That's all right if the clip art is an integral part of the transparency's content—such as the illustration of a computer network or the aerodynamic forces on an airplane.

Often, though, the clip art is no more than empty—and distracting—decoration:

The Five C's of Credit

- Character
- Capacity to repay
- Collateral
- Capital to cushion against losses
- Condition of the economy

There is a good place for creative clip art, however: the title transparency, which often is on the screen as the audience arrives. Because the presentation hasn't even started, the clip art can serve as visual interest.

Here's an example of effective clip art on a title transparency:

CINEMATIC TECHNIQUES

by

Liz Bailey

A creative title transparency is often better than one that uses only words:

CINEMATIC TECHNIQUES

by

Liz Bailey

We've looked at lots of tips for creating good visual aids. They help deliver your message and add pizzazz to your presentation. But there's a way to further reinforce your message and add even more pizzazz: involve the audience. That's the subject for the next chapter.

CHAPTER 8

Involving the Audience and Using Humor

Too many presentations seem to have an invisible separator between speakers and audiences. It's like a glass wall: the speakers stay on their side and do the talking; the audiences stay on their side and do the listening. (Theater people call that the "fourth wall.")

Good speakers usually try to eliminate that wall. One way is to let the audience participate during the presentation—people love to be part of the action. Another is to get them to laugh so they will feel greater rapport with the speaker. Either way breaks down the artificial separation that usually does more harm than good.

Let's start with involving the audience.

Involving the audience

There are various degrees of involving the audience. Some speakers have the audience members actually on their feet and moving around. Others simply ask for them to participate passively. Not all ways are appropriate for all presentations, but audiences are usually more willing to be active than inexperienced speakers assume.

We'll begin with the simplest way of making the audience more attentive.

Ask questions

Real questions invite audience participation. Unreal questions don't. An unreal question is when a speaker says something with a question mark at the end but doesn't pause for an answer: "Have any of you been to California recently? Well, I have. In fact. ... ." There's nothing wrong with an unreal question; it just doesn't create audience participation.

But a real question does. Here are some examples:

- An accountant, speaking on technical matters relating to income tax, started with a quiz. This was one of her questions: "Suppose my sister decides to separate from her husband, so she throws the bum out on July 4th. Which filing status can she use?" Excellent use of the question. If she'd started, instead, with a dry listing of filing statuses and the requirements for each, the audience would surely have moved deeply into the passive mode.

- An appliance salesman asked how many people in the audience owned microwave ovens. He counted the hands and then compared the result with the national average. This was an interesting way of telling us what could have been a dry statistic: 70% of the population owns microwaves.

- A writing consultant started his seminar by asking members of the audience what they thought of the writing in their in-boxes. He went around the room, getting a response from everyone. That way, he found out what the audience members were thinking and got them actively involved in his topic.

Notice these were all questions the audience could actually answer—often easily. When you ask questions—real ones—the glass wall disappears almost immediately.

Tip

Be sure to focus your questions. One speaker held out a 20-year-old swimming fin, asking a person in the audience to touch it and give her reaction. He asked, "What do you notice about this?" She said, "It seems flexible." Unfortunately, he'd intended for her to notice how heavy it was. He should have asked, "Does this seem heavy to you?"

Use the names of people in the audience

If you refer directly to people in the audience, you show you're willing for the glass wall to come down. If you know someone, try to single that person out: "I was impressed by Betsy's work on the reorganization committee. It will dramatically improve our company's. . . ." This creates a more informal atmosphere.

You can also use real people's names in scenarios you're creating. One speaker told a group about customer service representatives for a telephone company—and when they must refer questions to a supervisor. Here's how she used a person's name: "Poor Gene. The phone truck ran over his cat and killed it. He's calling me, a customer service rep, to complain. That's one I'd refer to my supervisor." Gene (who didn't even have a cat and was a simple prop) certainly straightened up at that point.

Later in the presentation, the speaker used the name of another person in the audience. This time she explained a telephone feature (Caller ID) that lets us know who is calling before we pick up the phone: "Remember Gera's first boy-friend? She was waiting by the phone, just hoping he'd call. She wanted to sound just right when the call came. Wouldn't Caller ID have been nice for her to have had?"

You can see that simply using real people's names can brighten up the audience. Gene and Gera—unaware they'd be named in the fictitious situations—took notice. And so did the rest of the audience, everybody glancing at the person the speaker referred to.

The speaker had another nice technique: as she spoke the person's name, she walked over and looked into the person's eyes.

Tip

Mention the names of earlier speakers—that shows you can be spontaneous. For example, in a session on buying houses, a lawyer mentioned the name of a real estate agent who had just spoken: "Let's say you've bought a house from Laura. Now you'll need some help from a lawyer. Let me tell you about the legal questions you'll need to be aware of."

Set up tasks

You can also involve people in more active ways in your presentation. I've seen speakers ask everyone in the audience to do these:

- Write our names with our non-preferred hand (to show how important it is for us to accept and emphasize our preferences in other matters).

- Communicate a sentence in sign language.

- Scrub for a medical operation (with pretend soap and water)—the procedure is amazingly thorough and complex.

- Do a breathing exercise designed to relieve stress (the speaker, in particular, benefited from this one).

Do these sound like they're not appropriate for business situations? For some situations, they're not. But business is exceptionally diverse, and these speakers were very successful.

Use members of the audience to demonstrate something

A demonstration involving volunteers really gets the audience's collective cardiovascular system going: those participating are fully alert; those not participating are still calming down from possibly being up front themselves; and everybody enjoys the action.

Tip

How do you get somebody to volunteer? In some cases, ask for help in advance. Often that's not possible. So just wait quietly and look at the audience. Someone always volunteers!

Here are some demonstrations I've seen:

- *Topic:* how to deal with an attack by muggers. The speaker had two volunteers demonstrate a mugging, one as the mugger, one as the victim. But first she explained what would happen, making especially sure the victim wouldn't overreact at being "attacked." This demonstration was extremely effective. Everyone in the audience could feel the violation of a personal attack.

- *Topic:* setting priorities. The speaker gave two members five seconds to grab as many poker chips as possible from a table. He later explained that the chips had different values: the blue were worth the most, then the red, then the white. The volunteers had to count the value of the chips they'd grabbed. The speaker then said that priorities at work have different values, too, and too often managers are so caught up in white-chip activities that they never get to the red-chip and blue-chip ones.

- *Topic:* easy ways to seal bottles. The speaker had a volunteer come forward and, using a simple device, put a cap *on* a bottle. This made the point clearly that the device was easy to use.

You can see the value of these exercises. The chairman of the board might not use them when talking to stockholders. But most presentations have the opportunity for some type of audience participation—from a relatively passive question to an active demonstration. Look for opportunities. They're often a presentation's highlight.

Tip

Be very, very sure you NEVER embarrass or trick your volunteers. They won't like it—and the audience won't either. And offer your thanks as the volunteers take their seats.

Using humor

Everybody likes humor. But not everybody agrees what's funny.

I've spoken many times when audience members filled out critique forms at the end. Early in my speaking career, I used to be surprised that jokes I felt were totally inoffensive had offended someone.

Soon I realized that the best humor is often unplanned—the little asides that happen spontaneously during a presentation, the kind that aren't necessarily funny in the cold light of day, but that provoke laughter during the heightened atmosphere of a presentation.

Other types of humor have their place. Let me give you a couple of examples.

A sailor recently gave an extremely funny presentation, defining nautical terms for non-sailors. I jotted down one of his comments: "If you're on my boat, don't say there's something on the *bow* of the boat. Just say there's something hanging off the *pointy end*—looks like a person. I'll know what you mean."

He then told us some of the most important "nautical terms" non-sailors need to know. This was his visual aid:

Nautical Terms To Know

- "Let go of _____ !" (*this, that, them*, etc.)

- "Pull this (or that)!"

- "DUCK!!!!"   (most important term)

This is inoffensive humor that everyone can (and did) enjoy.

Tip

If you expect to be nervous at the beginning of a presentation, avoid starting with narrative jokes. They require you to be at your best, and your timing must be perfect.

Sometimes a standard joke can work well, too. But usually the speaker should use himself as the "victim." For example, at a very large conference, the keynote speaker (a man at the top of his profession) started with this story:

> This morning, a woman came up to me and began a polite conversation:
>
> "Are you the speaker for today?" she asked.
>
> "Why, yes," I replied.
>
> "You must be quite nervous," she continued.
>
> "Well, not really. Why do you ask?" I said.
>
> "Then why else are you in the Ladies Room?"

Tip

Make yourself the target of humor. Nobody is going to be offended if you're the target, and jokes on yourself don't diminish your stature.

Humor spices up a presentation. Some you can plan; some just happens when you're especially alert and simply enjoying being the speaker.

As you can see, you can reinforce your message and add pizzazz by involving your audience and using humor. In fact, you should actually design these features into your presentation.

Once the design of your presentation is complete, you're ready to test it. That's what the next chapter is about.

CHAPTER 9

Rehearsing

People who design cars wouldn't dream of trying to sell them without testing them first. People who design airplanes wouldn't think of putting passengers on them without testing them first. People acting in plays wouldn't appear on opening night without a number of rehearsals.

But people who design presentations—or prepare them, at least—often run the first test in front of their real audience. That's not the time to find out the car won't start or the plane can't get off the ground or the five-act play won't reach the end of Act 1.

Rehearsal is a crucial part of preparing. Good speakers have usually said the words and tried the visual aids many times before standing up for real.

Rehearsing by yourself

You'll probably want your first rehearsal to take place by yourself. Simply go into a room that has the necessary equipment. Then give your presentation out loud (not just in your mind) to the wall and chairs. Again, again, and again.

When my daughter was young, she loved *The Little House on the Prairie* series of books. I once saw her finish a book and turn right to page one and start again. That's like what I do when I rehearse a new presentation.

Tip

> *Especially rehearse the beginning and the end of your presentation—the moments of greatest impact.*

One reason I rehearse is to find the words. My feeling is that if I can find the right words once, they'll come to mind more easily the next time. And the next. To me, that's much more effective than memorizing.

That reminds me of a line in E. B. White's famous essay, "Once More to the Lake." In it, White describes returning to his childhood vacation place decades later. He says, "It is strange how much you can remember about places like that *once you allow your mind to return into the grooves that lead back.*"

I think of rehearsing as making those grooves, of getting the words right, not so I can return to them decades later—but just a few days later. That way, when the ideas come to me, the words will, too.

Tip

> *If possible, rehearse at least once in the room you'll actually use for your presentation. You'll feel more like the room is "yours." Also, you'll find out about lighting, the on/off switch for the overhead projector, etc.*

When you rehearse, be sure to actually use your visual aids. You'll get practice introducing them properly and handling them. Just as important, you'll find out which ones don't project well and need work.

Rehearsing in front of others

Rehearsing by yourself helps you get the words right, but it doesn't really test your presentation. For that, you need other people. Unfortunately, the speakers who most need a live test audience are the ones least likely to ask for one. In fact, some speakers are more self-conscious rehearsing in front of a handful of co-workers than giving the presentation to the real audience.

Once you break through that self-consciousness, though, you'll find this is the most important way of all to rehearse. You'll find out what works and doesn't work with everything about your presentation—from the organization to the content to the visual aids to the audience involvement. The changes you make as a result will often be invaluable.

I suggest you try to get people to hear you who may be typical of your real audience. That way, you'll be running a true test.

When you do rehearse in front of people, ask them to not interrupt you the first time through—to take notes, instead. Otherwise, you'll be getting constant interruptions. Both you and your test audience will lose the flow of your presentation.

Tip

Have someone time your rehearsal. Audiences get restless if you go over your allotted time. And you'll get nervous if you fall way short.

In addition, I try to rehearse at least once in front of my wife. Other people tell me their spouses give pretty straightforward criticism, too.

Rehearsing in front of a camera

Many companies have videotape equipment. If they don't, you possibly have a video camera at home. Videotaping doesn't show you much about the organization or content of your presentation, but it does reveal a lot about your techniques of delivery: Are you pacing too much? Jittery? Saying "uh" or "okay"?

You don't need to videotape yourself more than once, but that once can let *you* see what your *audience* will surely see. And seeing it for yourself is much more effective than having other people tell you.

Tip
_____

When you're being videotaped, be sure the camera stays on you—not on the screen showing your visual aids. The intent is to videotape YOU, *not your presentation.*
_____

The next part of the book offers suggestions on how to give your presentation, covering such topics as setting up the room, presenting your visual aids, and using effective techniques of delivery. You'll want to read that part before you actually begin rehearsing.

Appendix A, "Checklist for Speakers," gives you reminders of what makes for a good presentation. You might want your test audience to have this checklist during your rehearsal.

GIVING YOUR
PRESENTATION

CHAPTER 10

Setting Up the Room

Even if your presentation is well designed and rehearsed, it may not succeed if the room isn't right. Little things—like a few extra degrees of temperature—can create a negative environment that makes listening almost impossible.

This chapter will cover what I look for in a room. I've learned most of these from hard experience. Here is what I check:

- the projector and screen
- the tables and chairs
- the microphone
- the lighting
- the temperature
- outside distractions

You may think an audiovisual crew (if there is one) should have all these ready, but from my experience, they don't. They arrange things from their perspective, but not from the perspective of an experienced speaker. Out of about a hundred presentations I give a year, the room is just right only a handful of times.

You may also think you don't have control of the room. Sometimes that's true. But most people have more control than they think. Get there early, let people know what you need, and they'll almost always be accommodating. The effort can make all the difference.

So let's look more closely now at these factors, and I'll give you some practical suggestions for each.

Checking the projector and screen

I can't rest until I know the audience can see my visual aids, so I go straight for the overhead projector. If I'm speaking in a hotel and get there the night before, I ask someone to open the room for me right then so I can check that projector. Otherwise, I won't be able to sleep.

This is what I look for:

- *Does the projector focus well?* Sometimes a projector focuses on part of the screen but not the rest. If that's the case, look for a knob to adjust the focal length. If there isn't one, you'll want to get another projector.

  If that isn't possible, try to get the best focus at the *top* of the screen and work as much as possible there. If you've designed your transparencies so most of the content is at the top, you'll at least make the best of a bad situation.

  By the way, I use a test transparency (see Chapter 6) for checking the overhead projector.

- *Is there a spare bulb—and does it work?* Normally turn off the projector before moving the lever on the projector to the spare bulb. Sometimes the spare is brighter than the original. In that case, use the spare.

- *When the transparency is straight on the face of the projector, does it project straight on the screen?* Projectors are hard to line up with screens, and they get hard use, too (things get bent). Often what looks straight on the face of the projector is quite crooked on the screen.

Sometimes you can move the projector left or right and get a different angle for the projection. That may solve the problem. If not, experiment until you find what angle you should use for placing the transparency on the projector, and remember that angle. Even better, use a strip of drafting tape across the face of the projector (near the top) as a guide.

- *Is the projected image large enough for the audience to see?* Go to the far corners of the room and look at the image. If you have good eyesight, make allowances for those who don't.

  If the image is too small, move the projector and screen farther apart so the projected image fills the screen as much as possible. If the image is still too small, consider rearranging the audience's tables and chairs (more later on that). Or get a larger screen.

- *Is the extension cord from the projector a hazard?* Because I can trip over anything, I try to avoid setting traps for myself. I normally stand with the projector on my right side as I face the audience, so I want the extension cord to run to an outlet even farther to the right. (Left-handed people should adjust accordingly.)

  I don't want the cord to run across any path I'll normally take during my presentation, except for occasional movements to the right side of the room.

- *Is the overhead projector on an adequate table?* I try to avoid stands designed just for overhead projectors. Usually they hold the projector and nothing else. I won't have room for my transparencies, and I need room not only for the stack of transparencies I'm going to use but also for the stack I've already used. The stands are also sometimes high, blocking the view of some members of the audience.

I try to find a moderately low table that has room for the projector and two stacks of transparencies. I'll leave room on the left side of the table for both stacks of my transparencies and set up the projector on the right.

Checking the tables and chairs

Once you have the equipment set up for your visual aids, you have to turn your attention to the seating for the audience. You usually don't have any control over the comfort of the chairs or the steadiness of the tables, but you probably do have control over their position.

- *Can the audience see all right?* If the seating is very close to the screen and very wide from left to right, people on the edges may not be able to see the screen. Their angle will be too shallow.

  If the seating goes too far back, people in the back may not be able to see. If necessary, rearrange the room (that's how I stay in shape). The local staff is normally quite helpful.

- *Can you move freely?* If you're like me, you want to move all around the room. For a small audience, I like the seating to be in the shape of a *U* with my table and projector filling part of the opening. Then I can move inside or outside the *U* as I speak.

Checking the microphone

A microphone is a great benefit if you have a large audience. If you're in the marginal area where you think the audience can "probably" hear you without one, ask for one anyway. Otherwise, you'll end up raising your voice unnaturally: that will strain your vocal cords (so you'll be hoarse the rest of the day); also, an unnaturally loud voice is hard for your audience to listen to—your natural inflection disappears.

You want a wireless microphone if at all possible. That way you can move freely to the projector, the screen, and the audience. A microphone with a wire is all right, but once again, that's a hazard for those of us who are clumsy.

Try to stay away from a fixed microphone—one that's part of a lectern. You'll be as immobile as the lectern.

If you are using a microphone, this is what you should check:

- *Is the volume adjusted correctly?* Talk over the microphone and have someone walk to all parts of the room to check the volume. When you talk, try to speak with the same loudness you expect to have during your presentation. In my case, I tend to speak just a bit louder than normal during a presentation, so I speak with a little extra loudness during the test, too.

- *Does it make unwanted noises?* I'm no electrical engineer, but audiovisual people tell me that a wireless mike needs a new battery very often. In fact, they'll normally replace the battery more than once a day during constant use. So if you hear a scratchy noise or static, have a new battery put in.

  If the microphone makes high-pitched squeals, lower the volume.

- *Can you turn your head to the side and still project?* Microphones have different designs. Some pick up your voice very well no matter how much you swivel your head from side to side. Others don't. This matters when you're turning your head toward the screen and talking at the same time: the microphone stays put, but the position of your mouth relative to it doesn't.

Tip

Don't put a lapel microphone on your lapel. It will be too far to one side or the other and sometimes not pick up your voice well. Instead, fasten it to your tie (for men) or to your blouse (for women).

- *Does it have a fastener that works?* Like projectors, microphones get hard use. And the fasteners for lapel mikes are often fragile. You don't want to find out there's a problem just as you're being introduced. If the fastener doesn't work, a safety pin can do wonders.

Tip

For women: Be careful of necklaces when you use a lapel microphone—sometimes they bang against it and make scraping or jangling noises.

Checking the lighting

You don't want a dark room. The lighting that makes your transparencies look perfectly beautiful is too dark. Your audience will begin to hibernate. Set the lights as bright as you can so the audience can still see the transparencies pretty well.

If you're using 35-millimeter slides or a computer presentation, you may need the lights a little lower than for transparencies. In that case, have someone lower the lights just when you use your visual aids. And have someone turn the lights back up when you're through with them.

Also, you don't want a light directly over your screen: it will shine on the screen and wash out your projected transparencies. If possible, move the screen a little so the light is behind it. If that won't work . . . well, I've had a lot of experience unscrewing light bulbs. Just get a steady chair.

Ideally, you want less light over the screen and nearly normal light elsewhere.

Checking the temperature

I hate being cold. But a warm room is a sleepy room. You want the temperature a little on the cool side when the audience is in the room. That means the room should be even cooler when it is empty. If the temperature seems just right when you're setting up, it will probably be too warm once everyone gets there.

Checking for outside distractions

Muzak. It sounds so nice you don't even notice it when you're setting up. But it becomes immediately obvious when the audience settles down and the introductions begin.

Another problem is the open door. You can't keep the door closed forever (the audience *will* come in). But if you hear outside noises even slightly during your presentation, the people sitting near the door are probably very distracted. Just ask someone to close it. People in the audience are usually glad to help.

Once you've set up rooms a few times, you'll learn what to look for. And when the room is the way you like it, you'll sense that it's *yours*. That helps you feel confident when the presentation begins.

Appendix B is a checklist for setting up the room.

CHAPTER 11

Using Effective Techniques of Delivery

Now your preparation is over, and you're ready for the actual presentation.

This is when speakers get nervous. Even though I've spoken many, many times, I occasionally feel the butterflies before an important presentation in front of a lot of people. That's natural. To calm myself down, I don't do yoga or deep breathing exercises. I simply look over my visual aids and review what I'm going to say for each one. Quickly I realize: "Oh, this is easy! I can do this!" I know my material, and the design and rehearsal have prepared me to present it.

I also try to think of my presentation as a conversation with my audience. I'll be the only one speaking much of the time, but, still, I'm just talking to them. Yes, I'll project more energy and have better focus, but I'll just be having a friendly conversation.

There are some techniques that can make that friendly conversation more effective. Here's what I recommend:

- start fast
- project energy

- move around the room
- make eye contact
- speak with good loudness and pace
- avoid distractions

Start fast

Most audiences decide pretty quickly whether they want to listen to you or not. Yet speakers who have great visual aids and a nice presentation sometimes ramble at the beginning—the time a visual aid isn't projected on the screen.

The solution? Say the minimum and get to your first visual aid. If you've designed your presentation well, it may be a funny quotation or something else that will quickly engage your audience.

An efficient start needn't seem abrupt to your audience. More likely, it will just seem like you're well prepared.

Tip

Here's a trick many speakers use to feel comfortable at the start: find a friendly face and talk to that person. You can create that friendly face by talking to members of the audience beforehand—especially if you're in the room as they come in. Then look at that person at the beginning of your presentation. As you gain confidence, look around. You should start finding more friendly faces there, too.

Project energy

Projecting energy doesn't necessarily mean moving freneti-cally. It simply means showing—through your posture, your movement, your facial expression, your voice—that you *care* about what you're saying. If you don't show that you care, how can you expect your audience to?

Sometimes, especially during a long presentation, you have to give yourself a pep talk, tell yourself to pick up your energy a notch. When you do, you'll almost invariably get the same response from your audience.

Projecting energy, by the way, doesn't mean doing anything out of the ordinary. Just act within your own personality: how do you behave when you really care about something? That's how you should behave during your presentation.

People who lack confidence in themselves are sometimes reluctant to project energy. They try to blend in with the wall. As a result, they fulfill their own expectations: they don't do as well as they should. So if you're not used to projecting energy in front of others, take a chance. You'll see the ben-efits immediately.

Move around the room

Someone is in control of the room. Either the speaker is or the audience is. When the audience is in control, speakers feel that the only space they own is where they're standing. When the speaker is in control of the room, audience mem-bers feel the only space they own is where they're sitting.

That sounds like a fight for control, doesn't it? But it really isn't. *Everyone* wants the speaker to be in control of the room.

And one of the best ways to show that control is to move around.

Movement is good for two other reasons:

- *It helps the audience.* Imagine people in the audience with their heads and necks in the same position for an entire presentation. That would be extremely uncomfortable, wouldn't it? But if you move around the room, you're giving them different angles, slightly different body positions, for viewing you.

- *It helps you.* Now imagine just standing in one place for your entire presentation. Could you stay there if you weren't giving a presentation? If so, what would the physical consequences be? Stiff legs, stiff back, stiff neck. So moving actually helps relax you as you speak.

Although movement is helpful, even crucial at times, I have seen successful presentations without it. I attended a presentation in an auditorium that held about 400 senior government executives. The speaker was sitting in a big easy chair on the stage. The audience was used to that: speakers always sat there during introductions.

The introduction took place, and the speaker never moved. He just sat there! The audience didn't know how to react because that had never happened before. Then the person in the chair started speaking. He was energetic. He leaned forward and fixed the audience with his eyes. Contrary to what we'd expect from someone just sitting in an easy chair, he was full of energy.

He didn't move around the room, but he took control in another way: he stayed in his chair. He did the unexpected. And he did everything else right. His delivery was so terrific that he didn't need to move around the room. He projected plenty of energy from where he was.

Tip

People wonder what to do with their hands while speaking. There isn't a simple answer. What do you do with them when talking to one or two people in an informal situation? That's what you should do with them when you're in front of many people. If you become conscious of your hands, don't put them in your pockets. If you're nervous, they'll just appear awkward. Instead, let them hang naturally by your sides. They won't be doing anything, but they won't be distracting, either. As you relax, your hands will naturally come alive again.

Make eye contact

Have you ever heard the suggestion to look just over the heads of the people in your audience? That's bad advice. You need to look right at them.

Imagine someone talking only to you for a few minutes without ever making eye contact: first that person looks down, then to the right, then to the left, then down again, all the while talking. That would seem pretty strange, wouldn't it? That's what happens when a speaker doesn't look at the audience. The audience starts to feel the speaker is talking *at* them, not *to* them.

Eye contact, by the way, isn't a behavior you try to acquire as a speaker. It's more a symptom—a symptom that you are actually trying to talk to people. When you're really trying to communicate, the eye contact is automatically there. It remains on an individual for at least a couple of seconds and then moves to another. It takes in all parts of the room.

Sometimes eye contact is hard when you're working at the overhead projector. The light from it is so bright that your eyes can't adjust to it and to the audience. In that case, you have to fake eye contact. You're actually "blind," but face outward and look different directions. You can still see vague shapes out there.

Tip
_____

> *Don't forget to smile (unless you're giving bad news or being stern on purpose). A smile really helps break the tension. There's a saying among speakers that "the face you give is the face you get." If you see lots of frowns in the audience, check your own expression.*

_____

Speak with good loudness and pace

If you speak too quietly for people to hear, you won't communicate. That's obvious. So if you're naturally soft-spoken, move closer to the audience and ask people to let you know if they have trouble hearing. Try to get a microphone if that's at all appropriate. Soft-spoken people using a microphone often have marvelous inflection that others didn't know was there.

Also, all speakers should try to speak fairly quickly. Lots of people think they speak *too* quickly. Only a few of them are right. Audiences much prefer a speaker who goes fast rather than one who goes too slow. Very slow speakers drive audiences crazy for a minute or two. Then most audiences tune them out.

Tip

Even though you should speak quickly, try to still have numerous pauses for emphasis. Think of pauses as punctuation. You'd have trouble reading a book with no punctuation, wouldn't you? Similarly, audiences have trouble listening to someone speak without pauses. There just aren't enough clues about what deserves emphasis.

Avoid distractions

If you're doing something distracting, the audience will have great difficulty paying attention to what you're saying.

You can distract people with verbal mannerisms: "uh" and "okay" are two of the most common (we've all counted other people's "uh's"). You cannot eliminate these entirely and shouldn't try. Just watch the top newscasters on television when they're not reading the news—when they're part of a round-table discussion or interviewing someone. You'll hear the "uh's," but probably not to distraction. And that's the key: keeping the verbal mannerisms under control so they don't distract your audience.

You can also distract with excessive movement. As I mentioned earlier, some movement is good and contributes to the energy of your presentation. That's purposeful movement. Movement that isn't so good is any jitteriness or noticeable pattern.

Some speakers pace back and forth, back and forth, back and. . . . That drives audiences crazy. Others wave the pointer wildly or click pens. To see if you have any distracting mannerisms, ask people to note them during your rehearsals. Or, even better, note them yourself by having your rehearsal videotaped.

There's another consideration for delivering your presentation: handling visual aids. There are some definite do's and don't's that can make or break it. They're the subject of the next chapter.

Appendix A, "Checklist for Speakers," is a quick summary of the main points of this chapter.

CHAPTER 12

Presenting Visual Aids

One of the most important techniques of delivery is presenting visual aids properly. That's why I've reserved an entire chapter for it. I'll concentrate on the overhead transparency —by far the most common type of visual aid. But, as I mentioned in Chapter 6, what works well for transparencies generally works well for other types of visual aids, too. The concepts are the same.

These are the two most important fundamentals in presenting visual aids:

- Don't block your audience's view.
- Direct your audience's attention where you want it.

The rest of the chapter will discuss these in more detail.

Don't block your audience's view

Usually the overhead projector is between the audience and the screen. So whenever you're standing by the projector, *you're* between the audience and the screen, too—often blocking the view. Here's an illustration of what I mean:

As you can see, when you're standing by the projector, you're in the wrong place: you're blocking the view of people on your left.

To keep out of their way, put a transparency on the projector and *then move immediately back to the screen.* Here's an illustration of the right place to stand:

This seems obvious, but beginning speakers almost always stand in the wrong place. They have to be by the projector to put transparencies on and take them off. Then they stay there—in the one place they must be during a presentation—and never move.

Tip

> *Be sure the area next to the screen is clear. You don't want to worry about a chair there or a table or an extension cord to trip over.*

Giving the audience a clear view of the screen requires conscious effort on your part. Moving to the screen has a side benefit, too: it also moves the audience's attention there. That's what the next section is about.

Direct your audience's attention where you want it

Good speakers are conscious of directing the audience's attention. Sometimes you want members of your audience to look at the screen; other times you want them to look at you. It's up to you to help the audience look the right place.

Most good speakers believe the audience can have its conscious attention only one place at a time. For example, if the speaker is talking and there's also a transparency projecting on the screen, what should the members of the audience do: Should they listen to the speaker? Or should they study the new transparency?

If they try to do both, they'll probably do neither. They can't pay attention to the speaker because there's a transparency on the screen. And they can't pay attention to the transparency because the speaker is talking.

So here are some suggestions for helping the audience keep its attention where you want it:

- Show the visual aid at the right time.
- Remove the visual aid at the right time.
- Cover parts of some visual aids.
- Use a pointer.
- Read most visual aids aloud.

Let's look more closely at these.

Show the visual aid at the right time

The instant you show a transparency, the audience will look at it. You can take advantage of that natural reaction by showing the transparency only when you want the audience to look at it.

Too often, though, beginning speakers confuse and distract audiences by doing things in the wrong order:

- First they show the transparency to the audience.
- Then they introduce it.
- Then they comment on it in detail.

You can see the problem—members of the audience look at the transparency while the speaker is introducing it but not talking specifically about what's on it. Their attention is divided—seeing one thing and hearing another.

My suggestion is to use this order instead:

- First, introduce the transparency (without showing it).

- *Then* show it to the audience.
- And then—immediately—comment on it in detail.

That way, the audience won't have its attention divided.

If you're handling your own transparencies, you won't have any trouble knowing what your next transparency is without showing it to the audience. Just look down at your stack of transparencies and read the label of the top one. You can tell what's next, but the audience can't.

A problem can arise, though, if someone else handles your transparencies while you speak: he or she has the stack; you don't. How can you tell what your next transparency is if you can't see it? That's one good reason to handle your own transparencies (many top executives do).

Tip

If you have a big stack of transparencies, don't put them all on your table at once. Just put enough to last until the next break. That way the audience won't get worried about the number of transparencies remaining—and the length of time your presentation will take.

If you must have someone else handle your transparencies, keep a list of them handy as you speak. Computer graphics programs can produce such a list easily, even showing several miniature copies of transparencies on a single page. The next page shows an example.

By the way, this is also handy when you're designing your presentation. It helps you study the flow of the presentation and prepare your transitions.

Tasks of a Desk Editor
- pagefitting
- stylizing
- sourcing

Pagefitting
- placing the text on the page
- changing the text to fit

Stylizing
- external
- internal

Sourcing
- checking facts
- checking MORE facts

Speaker's Notes

Remove the visual aid at the right time

So the right time to *show* the transparency is the instant you're ready for the audience to look at it. Does that mean the right time to *remove* the transparency is the instant you're through with it? Not necessarily.

If the transparency is simple—perhaps containing only a few bullets—it won't be a distraction on the screen while you're introducing the next topic. Simply leave it there. By then the audience has probably seen enough of it that they're looking elsewhere for amusement.

But if the transparency is a graph or an interesting picture or a chart full of controversial data, remove it as soon as you're through. Turn off the projector if you have a long introduction to the next transparency. Or just leave the projector on (projecting only light) if your introduction is short.

Some people object mightily to having a projector on without a transparency showing. They feel the bright light distracts the audience. So they turn the projector off every time they introduce a transparency. Some even turn the projector off and on every time they change a transparency.

I don't think that's a good practice for two reasons: First, all that clicking on and off certainly distracts the audience. They'll start watching the mechanics of what the speaker is doing instead of listening. And second, the bulb in the projector is much more likely to burn out when the projector is being turned on. Constantly turning the projector on and off greatly increases the chance of blowing out the bulb in the middle of your presentation.

Cover parts of some visual aids

This suggestion is controversial. Some people hate it. Some people love it. I'm in the second category—a real believer.

Fairly often a speaker has a transparency with several related points on it, like this:

> ## Advantages of Concurrent Engineering
>
> • gives our customers a better product
>
> • reduces the size of our maintenance force
>
> • saves far more money than it costs

Once the speaker introduces it and shows it on the screen, he plans to talk a little about each point. Members of the audience, however, aren't likely to look at only the first point while the speaker is talking about it. Instead, they'll naturally let their attention wander to point two, then to point three, then think a little about point two, then wander back to. . . . All the while, the speaker is still talking about point one.

The problem? The audience's attention is divided, and the speaker unintentionally provided the distraction (points two and three).

One solution is to have three transparencies with a point on each one. That works sometimes. But other times, the speaker wants all three on the screen at the same time. Or has only quick comments about each point and doesn't want the distraction of quickly moving transparencies on and off the projector.

In that case, cover the second and third points while talking about the first one. A sheet of paper works fine (or a system of overlays). As you finish talking about the first point, simply slide the cover down so points one and two are visible. And so forth. Here's an illustration of how I use a piece of paper to cover the second and third points while talking about the first one:

Advantages of Concurrent Engineering

• gives our customers a better product

When I'm ready for point two, I simply slide the paper down—like the illustration on the next page.

By the way, an advantage of using a piece of paper for the cover is that you can read through it when you look at the overhead projector. In other words, you can see what your next point is (so you can introduce it), but your audience can't. Another advantage of the cover is that you can write brief notes on it.

So try using a cover. You'll clearly see how powerfully you can focus your audience's attention. I use a cover constantly in all my briefings. Once you try it, you'll feel lost without that piece of paper.

Advantages of Concurrent Engineering

- gives our customers a better product

- reduces the size of our maintenance force

Use a pointer

A pointer, used correctly, is an extremely effective tool. For example, picture a speaker who only stands beside the screen and talks—without using a pointer to direct the audience's attention:

Now picture another speaker who stands beside the screen and uses a pointer each time she wants the audience to look there:

See the difference? The second speaker constantly directs the attention of members of the audience—letting them know when to look at her and when to look at the screen. Even during a short presentation, members of the audience can relax, letting the speaker do the work of directing their attention.

The pointer has another advantage: it not only directs the attention of the audience; it also visually reinforces transitions. We've all had the experience of speakers saying, "Now for my fourth point . . ."—and we thought they were still on their first one! But the pointer unmistakably takes the audience from point one to point two, from point two to point three, and so forth.

A third advantage of using the pointer is that it creates purposeful movement for you and for your hands. You're doing something controlled and professional looking. You don't have to worry any longer about what to do with your hands during a presentation. And the pointer moves you to the screen—the right place to be.

Some people point by placing a pen on the transparency itself. However, that tends to anchor the speaker to the projector, interfering with the audience's view. I recommend standing by the screen and using a simple wooden pointer or the metal kind that looks like a telescoping antenna. Just be watchful of fiddling with it and distracting the audience. Set the pointer aside when you're not using it.

By the way, I suggest avoiding "laser" pointers—the kind that project a light (often an arrow) on the screen. If you're like me, you won't be able to hold the light steadily on one spot, so the audience will see an arrow wobbling crazily on the screen. And there's a great temptation to move the light all over the screen while talking (the same thing can happen during computer presentations when speakers use the mouse arrow as a pointer—lots of meandering and jittery movement).

If you really want to use a laser pointer, just flash it momentarily on the screen and then turn it off. That gives the value of a pointer and minimizes distraction.

Tip

When you use a pointer, face the audience and hold the pointer in your hand closer to the screen. If you use the hand away from the screen, you tend to turn your body away from the audience and toward the screen. That can make you harder to hear. It can also draw you more in front of the screen so you block the audience's view.

Read most visual aids aloud

Audiences hate having speakers do nothing but read to them—no talking, no side comments, just reading one transparency after another. And I have to agree.

But there's valuable reinforcement for an audience to *see* the words on a screen and *hear* them, too. So I suggest reading every word on a transparency aloud to your audience. Of course, as I mentioned in Chapter 6, there shouldn't be very many words on your transparency—normally just what would be the headings and subheadings if you'd actually written down your entire speech. In other words, your "paragraphs" aren't on the transparency, just the headings and subheadings.

Occasionally, though, you do need more than just headings and subheadings. Perhaps you have an especially good quotation. Or a precise definition. You want the audience to comprehend every word. Again, I suggest you actually read the words aloud.

From my experience, asking members of the audience to read the words to themselves doesn't work very well. Many of them accept that as an invitation for a mental vacation.

To keep the audience from feeling "read to," however, intersperse a few comments as you're reading. For example, suppose I want to read aloud a quote from my favorite writer, E. B. White, showing how masterfully he uses simple language. This would be my transparency:

E. B. White's First Girl Friend

Her name was Eileen. She was my age and she was a quiet, nice looking girl. She never came over to my yard to play, and I never went over there, and, considering that we lived so near each other, we were remarkably uncommunicative; nevertheless, she was the girl I singled out, at one point, to be of special interest to me. Being of special interest to me involved practically nothing on the girl's part—it simply meant that she was under constant surveillance. On my own part, it meant that I suffered an astonishing disintegration when I walked by her house, from embarrassment, fright, and the knowledge that I was in enchanted territory.

After the phrase "she was under constant surveillance," I'd probably say, "Remember those days?" The simple comment helps prevent the audience's perception that I'm only reading to them.

So keep in mind the two fundamentals of presenting visual aids: let members of the audience see the screen and keep their attention where you want it to be. The techniques require rehearsal but soon become automatic—just like driving a car with a stick shift.

Now let's turn our attention to the finale of your presentation: handling questions and answers.

CHAPTER 13

Handling Questions and Answers

Sometimes there isn't a formal question-and-answer session—members of the audience ask questions throughout. That's especially good for small, informal sessions. The questions help break the invisible "glass wall" between speaker and audience. They can also help relax the speaker by turning the presentation into a conversation.

For some presentations, however, questions may be interruptions. If someone does interrupt with a question, just give a quick, polite response (including the bottom line) and explain that you'll go into more detail later.

If you do intend to have a formal question-and-answer session, you can plan for it—you needn't be totally at the mercy of your audience. You can prepare so that the unexpected becomes the exception:

- One way is to imagine the questions you'll most probably be asked and prepare for them.

- Another way is to imagine the questions you'd *hate* the audience to ask—then prepare answers for them. I've seen speakers for important presentations brainstorm these questions and answers with key members of their staff. And then rehearse answering the questions. Politicians heading into debates do this regularly.

In either case, consider preparing visual aids to support you during the question-and-answer session. They have the same value here as they do in the regular part of your presentation.

However, even if you are confident of the content for a question-and-answer session and have some backup transparencies, there are still some techniques to consider. For the rest of the chapter, I'll cover:

- how to start the question-and-answer session
- where to stand and where to look
- how to handle difficult questioners
- how to end the session

How to start the question-and-answer session

Normally you'll finish your main presentation with a brief summary. The audience should know you're through by the content and the inflection in your voice.

To give another signal that the presentation is over, turn off the projector (or other electronic device, if you're using one) and move toward the audience. At the same time, specifically ask for questions: "That's the end of my formal presentation. I have ten minutes for questions. Does anybody have one?"

Sometimes audiences aren't immediately ready to ask questions—they simply haven't had time to think of any. If you sense there are genuine questions waiting but the audience just isn't prepared, you can ask the first question yourself: "Well, let me start. A common question is just how often. . . ."

When you're through with your own (brief) answer, ask again for questions. Pause and make significant eye contact. Just wait. A question is bound to happen. The silence will grow uncomfortable, and you can wait longer than they can.

On the other hand, there's nothing wrong with no questions whatsoever. If the members of the audience really don't have any, why create artificial questions?

Where to stand and where to look

A general rule is to move closer to the audience during the question-and-answer session. That gives a subconscious signal that you're confident by showing you'll enter into the audience's space with no visual aids, notes, or other props. In fact, if there's a stage and you have a portable microphone, consider actually leaving the stage and walking out into the audience.

If you're in a smaller room, you'll be tempted to move toward the person asking a question. That's fine unless the questioner has a soft voice. Then you're *really* tempted to move toward the questioner—but you shouldn't. Instead, move farther away. That will usually make the questioner speak more loudly so everyone can hear. And audiences are more comfortable hearing the actual question rather than a rephrasing by the speaker.

If the audience still can't hear the question, however, be sure to repeat it for everyone before answering.

Tip

If you don't know the answer, say so. The audience can always tell.

Start your answer by looking directly at the questioner. Then look outward and include the rest of the audience. Look back several times at the questioner to be sure you're answering properly. In general, spend about a third of the time looking at the questioner and two-thirds looking at the rest of the audience.

Tip

When you are through answering a question, normally ask the questioner if you answered it all right. If not, try again. Everybody in the audience appreciates a speaker who takes questions seriously.

If the questioner is an important person or you especially want to be sure you've answered the question well, finish by standing near the questioner and looking at him or her.

On the other hand, if you don't want to hear from that person again, you have a couple of choices:

- finish on the far side of the room from that person
- finish right next to that person but look outward at the rest of the audience

Then ask for more questions.

How to handle difficult questioners

You don't want the final impression of your presentation to be an awkward situation during the question-and-answer session. Here are three types of difficult questioners and some advice on how to handle them (or try to handle them, at least):

- *The long-winded questioner.* Some people seem more interested in giving speeches than in genuinely asking for information. That's normally presumptuous of them and impolite to the rest of the audience. The main way to end their "speech" is to break in at the tiniest pause, pretend that was the end of their question, and resume control of the room by giving your answer. Then, of course, don't ask if you've answered their question satisfactorily and don't ever look their way again.

- *The incomprehensible questioner.* Sometimes it's hard to figure out what the question is. When that happens, I assume the questioner has good intentions, so I try rephrasing the question: "I'm not sure I understand your question. Are you asking . . . ?" Work for a minute to get on the same wave length with the questioner. In a case like this, the audience appreciates your sincerity.

- *The hostile questioner.* Some questioners have impure motives, so there's little you can do to turn them into understanding, sympathetic friends. Your goal, normally, is to limit the damage. Answer briefly (and normally forcefully) and turn elsewhere.

  Above all, don't get "hooked"—that's when the speaker loses control and argues with the questioner. Or gets sarcastic. Or nasty. That's the surest way to lose not only your temper but the entire audience as well. Try to assume the demeanor of a calm, serious, unemotional, and well-intentioned person. Try.

How to end the session

The most common way to end the question-and-answer session is with a smile and a "thank you." But you can add some polish, too. A final transparency (I try to get something relevant and humorous) sends the audience out smiling and shows you've thought through your presentation. You add just a bit of formality at the end—a finishing touch.

FINAL WORDS

CHAPTER 14

Helping Others Speak Better

Perhaps you have people working for you who give presentations. Or there's a co-worker or friend who needs some help. How can you help them become better speakers?

Too often, those helping with a presentation simply look through the other person's transparencies and then offer suggestions. However, that's often too late in the process. And, because there's no rehearsal, it doesn't show what will happen when the person actually stands up.

The time for you to get involved is in the *design* stage—helping them design a presentation that will be easier to give. You should also propose a rehearsal or two. Beginning speakers are often nervous about rehearsing in front of others. So during the first rehearsal, I normally emphasize what they do well and what they need to work on to get the *design* of the presentation right.

Once the design is right, many of the nervous mannerisms and excessive "uh's" tend to disappear—they're often the product of not knowing what to say next. A good design often takes care of that problem.

At some point, though, you need to say what needs to be said—the problems with delivery that just aren't going away.

Most people genuinely appreciate finding out. Usually no one has told them before, and they're relieved to know exactly what to work on. However, I'm careful not to point out any problems the speaker can't change—such as an unusual-sounding voice.

You can also give feedback to people after their actual presentation. The best kind is positive if that's at all appropriate. Positive feedback can create a good cycle: people who have succeeded in a presentation gain confidence, and that helps them do better the next time.

Bring up the negatives much later when you're helping the person prepare the next presentation. Even then, I try to stay positive. For example, I don't normally say, "Last time you really screwed up the question-and-answer session because you weren't prepared." Instead, I'd suggest working with that person to prepare the qustion-and-answer session.

This sounds awfully gentle for a professional situation, but a lack of confidence is a crucial reason most presentations aren't as successful as they should be. Improve the confidence, and the speaking improves along with it.

Finally, one of the best ways to help other speakers is to have them help *you* prepare *your* presentation. That way they can see the full process for designing a good presentation. And they can see the value of rehearsing.

There are also a couple of suggestions for helping *you* speak better. Look carefully at other speakers and try to understand why their presentation is successful or just ho-hum.

If you notice someone really effective, ask yourself why. Perhaps there's something special—humor, audience participation, energy—that you'll want to emulate. And if you start to get sleepy during a presentation, try to figure out why. What's the speaker doing wrong that's causing you to lose attention?

In effect, you can continue to improve just sitting in the audience—and appreciate what makes most presentations work: an excellent *design*!

APPENDIXES

Checklist for Speakers

Organization

- Do you clearly announce your topic?
- Do you define key terms in your topic?
- Do you state your bottom line up front (unless your presentation is purely informative)?
- Do you give a blueprint (actual or implied) in your introduction?
- Do you have *strong* transitions beginning each major section?

Content

- Do you use frequent examples?

Voice

- Is your speaking free of distractions ("uh," "um," "you know")?
- Do you use natural, spoken inflection?
- Do you use natural pauses for emphasis?

- Do you use appropriate loudness: not too soft or too loud?
- Do you use appropriate pace: not too slow or too fast?

Movement

- Do you have *any* purposeful movement around the room?
- Is your movement natural and free of distractions (rocking, pacing, jitteriness, hands in pockets, etc.)?
- Do you use natural hand gestures?

Visual Aids

- Are your visual aids designed well (uncluttered, readable, short items rather than long paragraphs)?
- Do you reveal them at the right time—just when the audience needs to turn attention to them?
- Are they visible to everyone in the room (you don't block the screen)?
- Are they straight (for transparencies)?
- Do you use a pointer?

Overall

- Is your audience paying attention at all times?
- Does your audience appear to understand everything you say?
- Do you appear interested in your topic?
- Do you have good eye contact?
- Do you appear obviously rehearsed?

Checklist for Setting Up the Room

Projector and Screen

- Does the projector focus well?
- Is there a spare bulb?
- When the transparency is straight on the face of the projector, does it project straight on the screen?
- Is the projected image large enough for the audience to see?
- Is the extension cord from the projector a hazard?
- Is the overhead projector on a table large enough to hold the stack of transparencies you *will* use and the stack you *have* used?

Table and Chairs

- Can the audience see the screen clearly from all seats in the room?
- Can you move freely about?

Microphone

- Do you have a wireless microphone?
- Does it have a new battery?
- Is the volume adjusted correctly?
- Can you turn your head toward the screen and still project clearly?
- Does the microphone have a fastener that works?
- Have you removed badges, necklaces, etc., that can clank against the microphone?

Lighting

- Is the area above the screen unlighted?
- Is the rest of the room at nearly normal lighting?

Temperature

- Is the room slightly on the cool side—especially before the audience has entered the room?
- Do you know where the thermostat is and whether or not you can make adjustments to it?

Outside Distractions

- Can you hear any unwanted noises (such as piped–in music)?

Model for Organizing
Your Presentation

This appendix is a model that can be helpful as you plan your presentation. It shows you a very obvious organization—and most busy listeners prefer obviousness to subtlety. After all, if the organization is unmistakable, then even complex, technical content can be a little easier to understand.

Don't feel you must follow this model explicitly, but it's surprising how often it can work. For details on the model, see Chapter 2, "Organizing Your Presentation."

Model for Organizing Your Presentation

Introduction

> Announce your main topic.
> Define your key term in plain English (and consider giving an example).
> State your bottom line.
> Give your blueprint.

Section 1

> Begin with a strong transition.
> Announce your topic for this section.
> State your bottom line for this section (if appropriate).
> Then present the details—remembering the value of examples.

Section 2 (etc.)

> Begin with a strong transition.
> Announce your topic for this section.
> State your bottom line for this section (if appropriate).
> Then present the details—remembering the value of examples.

Conclusion

> Begin with a strong transition.
> Restate your main point (unless that would seem unnecessarily repetitious).
> Turn off the overhead projector or other equipment to signal the presentation is over.
> Ask for questions.

Index